Holt Mathematics

Chapter 4 Resource Book

HOLT, RINEHART AND WINSTON
A Harcourt Education Company
Orlando • Austin • New York • San Diego • London

Printed in the United States of America

ISBN 0-03-078219-8

3 4 5 170 09 08 07

CONTENTS

Blackline Masters

Parent Letter	1
Lesson 4-1 Practice A, B, C	3
Lesson 4-1 Reteach	6
Lesson 4-1 Challenge	8
Lesson 4-1 Problem Solving	9
Lesson 4-1 Reading Strategies	10
Lesson 4-1 Puzzles, Twisters & Teasers	11
Lesson 4-2 Practice A, B, C	12
Lesson 4-2 Reteach	15
Lesson 4-2 Challenge	16
Lesson 4-2 Problem Solving	17
Lesson 4-2 Reading Strategies	18
Lesson 4-2 Puzzles, Twisters & Teasers	19
Lesson 4-3 Practice A, B, C	20
Lesson 4-3 Reteach	23
Lesson 4-3 Challenge	24
Lesson 4-3 Problem Solving	25
Lesson 4-3 Reading Strategies	26
Lesson 4-3 Puzzles, Twisters & Teasers	27
Lesson 4-4 Practice A, B, C	28
Lesson 4-4 Reteach	31
Lesson 4-4 Challenge	33
Lesson 4-4 Problem Solving	34
Lesson 4-4 Reading Strategies	35
Lesson 4-4 Puzzles, Twisters, & Teasers	36
Lesson 4-5 Practice A, B, C	37
Lesson 4-5 Reteach	40
Lesson 4-5 Challenge	41
Lesson 4-5 Problem Solving	42
Lesson 4-5 Reading Strategies	43
Lesson 4-5 Puzzles, Twisters & Teasers	44
Lesson 4-6 Practice A, B, C	45
Lesson 4-6 Reteach	48
Lesson 4-6 Challenge	49
Lesson 4-6 Problem Solving	50
Lesson 4-6 Reading Strategies	51
Lesson 4-6 Puzzles, Twisters & Teasers	52
Lesson 4-7 Practice A, B, C	53
Lesson 4-7 Reteach	56
Lesson 4-7 Challenge	57
Lesson 4-7 Problem Solving	58
Lesson 4-7 Reading Strategies	59
Lesson 4-7 Puzzles, Twisters & Teasers	60
Lesson 4-8 Practice A, B, C	61
Lesson 4-8 Reteach	64
Lesson 4-8 Challenge	65
Lesson 4-8 Problem Solving	66
Lesson 4-8 Reading Strategies	67
Lesson 4-8 Puzzles, Twisters & Teasers	68
Lesson 4-9 Practice A, B, C	69
Lesson 4-9 Reteach	72
Lesson 4-9 Challenge	73
Lesson 4-9 Problem Solving	74
Lesson 4-9 Reading Strategies	75
Lesson 4-9 Puzzles, Twisters & Teasers	76
Answers to Blackline Masters	77

Date ___________

Dear Family,

In this chapter, your child will learn about divisibility. Your child will also begin to understand fractions by comparing and ordering them and will receive an introduction to fraction operations.

As part of number theory, your child will learn that a number is **divisible** by another number if the quotient is a whole number with no remainder. There are rules to determine the divisibility of a number.

Divisibility Rules		
A number is divisible by	**Divisible**	**Not Divisible**
2 if the last digit is even (0, 2, 4, 6, or 8)	3,978	4,975
3 if the sum of the digits is divisible by 3	315	139
4 if the last two digits form a number divisible by 4	8,512	7,518
5 if the last digit is 0 or 5	14,975	10,978
6 if the number is divisible by both 2 and 3	48	20
9 if the sum of the digits is divisible by 9	711	93
10 if the last digit is 0	15,990	10,536

Your child will use this knowledge about divisibility to find the **greatest common factor** of a set of numbers.

Factors of 24: 1, 2, 3, 4, 6, 8, 12, 24
Factors of 36: 1, 2, 3, 4, 6, 9, 12, 18, 36
Common factors: 1, 2, 3, 4, 6, (12)
The greatest common factor of 24 and 36 is 12.

For understanding of fractions, your child will learn how decimals and fractions can be used to represent the same number.

A **decimal** can be written as a fraction: **0.23.**
Identify the place value of the digit farthest to the right in 0.23.

The **3** is in the **hundredths** place, so use **100** as the denominator. $\frac{23}{100}$

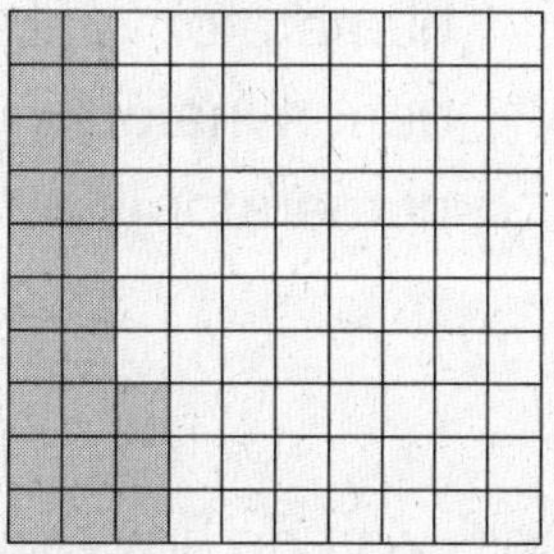

Your child will expand an understanding of fractions by learning about types of fractions and mixed numbers. **Mixed numbers** contain both a whole number greater than 0 and a fraction such as $\frac{1}{2}$.

A decimal can also be written as a mixed number. The number **2.7** can be written as $2\frac{7}{10}$.

Equivalent fractions are fractions that represent the same value. You can use multiplication and division to find equivalent fractions.

$$\frac{2}{3} = \frac{?}{18}$$

$$\frac{2 \times 6}{3 \times 6} = \frac{12}{18}$$

In the denominator, multiply 3 by **6** to get **18**. Multiply the numerator by the same number, **6**, to get **12**. So $\frac{2}{3}$ is equivalent to $\frac{12}{18}$.

A **proper fraction** is a fraction in which the numerator is less than the denominator.

$$\frac{2}{5} \qquad \frac{102}{351} \qquad \frac{5}{6}$$

An **improper fraction** is a fraction in which the numerator is greater than or equal to the denominator.

$$\frac{11}{4} \qquad \frac{3}{3} \qquad \frac{222}{115}$$

Your child will learn to **add and subtract fractions** with like denominators.

$$\frac{1}{8} + \frac{1}{8}$$

$$\frac{1}{8} + \frac{1}{8} = \frac{2}{8}$$

$$= \frac{1}{4}$$

Add the numerators. Keep the same denominator. Using the knowledge of greatest common factor, you can write the answer in simplest form.

Your child will learn to estimate fraction sums and differences. First, estimate the fractions by rounding to 0, $\frac{1}{2}$ or 1.

$$\frac{5}{6} + \frac{7}{12}$$

$$1 + \frac{1}{2} = 1\frac{1}{2}$$

$\frac{5}{6} + \frac{7}{12}$ is about $1\frac{1}{2}$.

Think: $\frac{5}{6}$ rounds to 1 because the numerator is about the same as the denominator. $\frac{7}{12}$ rounds to $\frac{1}{2}$ because the numerator is about half the denominator.

For additional resources, visit go.hrw.com and enter the keyword MR7 Parent.

Name ______________________ Date __________ Class __________

LESSON 4-1

Practice A

Divisibility

Tell whether each number is even or odd.

1. 10 ________

2. 35 ________

3. 20 ________

4. 14 ________

5. 17 ________

6. 19 ________

Tell whether each number is divisible by 2, 3, 4, 5, 6, 9, and 10.

7. 32 ________

8. 81 ________

9. 45 ________

10. 56 ________

11. 21 ________

12. 72 ________

Tell whether each number is prime or composite.

13. 4 ________

14. 9 ________

15. 23 ________

16. 11 ________

17. 12 ________

18. 16 ________

19. Three brothers bought a pack of 24 baseball cards to share. Can they each get the same number of cards? How do you know?

____

20. Five friends decide to order a pizza. If the pizza is cut into 8 slices, will they each be able to get the same number of slices without any pizza left over? How do you know?

____

Name ______________________ Date __________ Class __________

LESSON 4-1

Practice B

Divisibility

Tell whether each number is divisible by 2, 3, 4, 5, 6, 9, and 10.

1. 90 ______

2. 416 ______

3. 308 ______

4. 540 ______

5. 804 ______

6. 225 ______

7. 663 ______

8. 972 ______

9. 836 ______

Tell whether each number is prime or composite.

10. 33 ______

11. 69 ______

12. 41 ______

13. 45 ______

14. 58 ______

15. 87 ______

16. 61 ______

17. 53 ______

18. 99 ______

19. Dan counted all the coins in his bank, and he had 72 quarters. Can he exchange the quarters for an even amount of dollar bills? How do you know?

______

______

20. A small town purchased 196 American flags for its Memorial Day parade. Eight locations were selected to display the flags. Can each location have the same number of flags? If no, explain why not. If yes, how many flags will be displayed at each location?

______

Name ________________ Date ________ Class ________

LESSON **4-1**

Practice C
Divisibility

Tell whether each number is divisible by 2, 3, 4, 5, 6, 9, and 10.

1. 1,524 ________

2. 1,000 ________

3. 4,455 ________

4. 2,160 ________

5. 3,174 ________

6. 8,433 ________

7. 5,745 ________

8. 6,200 ________

9. 13,000 ________

Write all the prime numbers between each set of numbers.

10. 1–10 ________

11. 21–30 ________

12. 35–45 ________

13. 50–60 ________

14. 65–75 ________

15. 75–100 ________

Replace each question mark with a digit that will make the number divisible by 9.

16. 35**?** ________

17. 2,68**?** ________

18. 7**?**4 ________

19. 53,**?**69 ________

20. Several Boy Scouts worked together to create a line of 4,900 dominoes. They each placed the same amount of dominoes in the line. There were between 50 and 100 Boy Scouts working on the line. What are all the possible numbers of Boy Scouts who could have created the domino line?

____

21. Use the divisibility rules you know to write a divisibility rule for 12.

____

____

Name ______________________ Date __________ Class __________

LESSON **4-1**

Reteach

Divisibility

A number is divisible by another number if the quotient is a whole number with no remainder.

15 is divisible by 3 because $15 \div 3 = 5$. There are rules to help you figure out if a number is divisible by another number.

Divisibility Rules

A number is divisible by:

2 if the last digit is even.
3 if the sum of the digits is divisible by 3.
4 if the last two digits form a number that is divisible by 4.
5 if the last digit is 0 or 5.
6 if the number is divisible by both 2 and 3.
9 if the sum of the digits is divisible by 9.
10 if the last digit is 0.

To tell whether 315 is divisible by 2, 3, 4, 5, 6, 9, and 10, you can use the divisibility rules listed above.

5 is not an even number, so 315 **is not divisible** by 2.
$3 + 1 + 5 = 9$. 9 is divisible by 3, so 315 **is divisible** by 3.
15 is not divisible by 4, so 315 **is not divisible** by 4.
The last digit is 0 or 5, so 315 **is divisible** by 5.
315 is divisible by 3, but not 2. So, 315 **is not divisible** by 6.
The sum of the digits is 9 and 9 is divisible by 9. So 315 **is divisible** by 9.
The last digit is not 0, so 315 **is not divisible** by 10.

315 is divisible by 3, 5, and 9.

Use divisibility rules to tell whether each number is divisible by 2, 3, 4, 5, 6, 9, and 10.

1. 120 ____________________

2. 435 ____________________

3. 228 ____________________

4. 540 ____________________

5. 144 ____________________

6. 634 ____________________

7. 402 ____________________

8. 320 ____________________

Name ________________________ Date ____________ Class ____________

LESSON 4-1

Reteach

Divisibility (continued)

Numbers that are divisible by more than 2 numbers are called composite numbers.

12 is a composite number because it is divisible by 1, 2, 3, 4, 6, and 12.

$12 \div 1 = 12$	$12 \div 2 = 6$	$12 \div 3 = 4$
$12 \div 4 = 3$	$12 \div 6 = 2$	$12 \div 12 = 1$

Numbers that are divisible by exactly 2 numbers, 1 and itself, are called prime numbers.

5 is a prime number because it is divisible by only 1 and 5.

$5 \div 1 = 5$ $\quad$ $5 \div 5 = 1$

Tell whether each number is prime or composite.

9. 7 ____________ **10.** 12 ____________ **11.** 17 ____________ **12.** 28 ____________

13. 37 ____________ **14.** 45 ____________ **15.** 79 ____________ **16.** 88 ____________

17. 72 ____________ **18.** 59 ____________ **19.** 94 ____________ **20.** 21 ____________

21. 47 ____________ **22.** 55 ____________ **23.** 67 ____________ **24.** 81 ____________

Name ______________________________ Date ____________ Class ____________

LESSON **4-1**

Challenge

The Chinese Calendar

The Chinese calendar runs in cycles of 12 years. Each year in a cycle is named after an animal. The Chinese believe the animal ruling the year in which you were born has a great effect on your personality. They say, "This is the animal that hides in your heart."

2000 was a year of the dragon.

The Chinese calendar cycle below is for the years 1983–1994. Use the clues to write the last two digits in the year that goes with each animal. Then read the character traits associated with the animals. Which animal hides in your heart?

Tiger: 19 ______ **Divisible by:** 2 and 43 **Traits:** bold, adventurous	**Sheep:** 19 ______ **Divisible by:** 7 **Traits:** charming, artistic	**Pig:** 19 ______ **Divisible by:** 1 and 83 **Traits:** loyal, tolerant
Dog: 19 ______ **Divisible by:** 2 and 47 **Traits:** honest, faithful	**Dragon:** 19 ______ **Divisible by:** 2, 4, and 8 **Traits:** fun, energetic	**Rat:** 19 ______ **Divisible by:** 4 and 6 **Traits:** generous, creative
Horse: 19 ______ **Divisible by:** 5 and 9 **Traits:** friendly, hardworking	**Rooster:** 19 ______ **Divisible by:** 3 and 31 **Traits:** decisive, proud	**Ox:** 19 ______ **Divisible by:** 5 **Traits:** confident, leaderlike
Snake: 19 ______ **Divisible by:** 1 and 89 **Traits:** charming, wise	**Monkey:** 19 ______ **Divisible by:** 2 and 4 **Traits:** smart, likable	**Rabbit:** 19 ______ **Divisible by:** 3 and 29 **Traits:** friendly, cooperative

Name ______________________ Date __________ Class __________

LESSON **4-1** **Problem Solving**
Divisibility

Use the table to answer the questions.

Subways Around the World

City, Country	Length (mi)
New York, U.S.	247
Mexico City, Mexico	111
Paris, France	125
Moscow, Russia	152
Seoul, South Korea	83
Tokyo, Japan	105

1. Which city's subway has a length that is a prime number of miles?

____

2. Which subway could be evenly broken into sections of 2 miles each?

____

3. Which subways could be evenly broken into sections of 5 miles each?

____

____

Circle the letter of the correct answer.

4. Which subway's length is divisible by 4 miles?

A New York, United States
B Paris, France
C Tokyo, Japan
D Moscow, Russia

5. Which subway's length is not a prime number, but is also not divisible by 2, 3, 4, 5, 6, or 9?

F Mexico City, Mexico
G New York, United States
H Seoul, South Korea
J Paris, France

6. The subway in Hong Kong, China, has a length that is a prime number of miles. Which of the following is its length?

A 260 miles
B 268 miles
C 269 miles
D 265 miles

7. The subway in St. Petersburg, Russia, has a length that is divisible by 3 miles. Which of the following is its length?

F 57 miles
G 56 miles
H 55 miles
J 58 miles

Name ______________________ Date ___________ Class ___________

LESSON 4-1

Reading Strategies

Use a Graphic Aid

1	2	3	4	5	6	7	8	9	10
11	12	13	14	15	16	17	18	19	20
21	22	23	24	25	26	27	28	29	30
31	32	33	34	35	36	37	38	39	40
41	42	43	44	45	46	47	48	49	50
51	52	53	54	55	56	57	58	59	60
61	62	63	64	65	66	67	68	69	70
71	72	73	74	75	76	77	78	79	80
81	82	83	84	85	86	87	88	89	90
91	92	93	94	95	96	97	98	99	100

Circle all the multiples of 2 on the hundred chart.

1. Write the first number you circled in the second row.

____

2. Divide 12 by 2. What is the answer?

____

Divisible means **"can be divided by."** A number is divisible by another number if the quotient is a whole number with no remainder. 12 can be evenly divided by 2. All multiples of 2 are divisible by 2.

Circle the multiples of 3, 5, and 10 on the chart in three different colors.

3. Is each number that is divisible by 10 also divisible by 5? Why or why not?

____

4. Is every number that is divisible by 5 also divisible by 10? Why or why not?

____

5. Is every number that is divisible by 5 also divisible by 3? Why or why not?

____

Name ______________________ Date ____________ Class ____________

LESSON **4-1**

Puzzles, Twisters & Teasers

Monkey's Web

Monkeys swinging among the trees created webs of tangled vines. There is an inner web and an outer web. It's your job to connect them so that the monkeys can run between them. Two points in the web can be connected by a straight line only if the number in the outer ring is evenly divisible by the number in the inner ring. Two points cannot be connected if the straight line would cross through the inner ring.

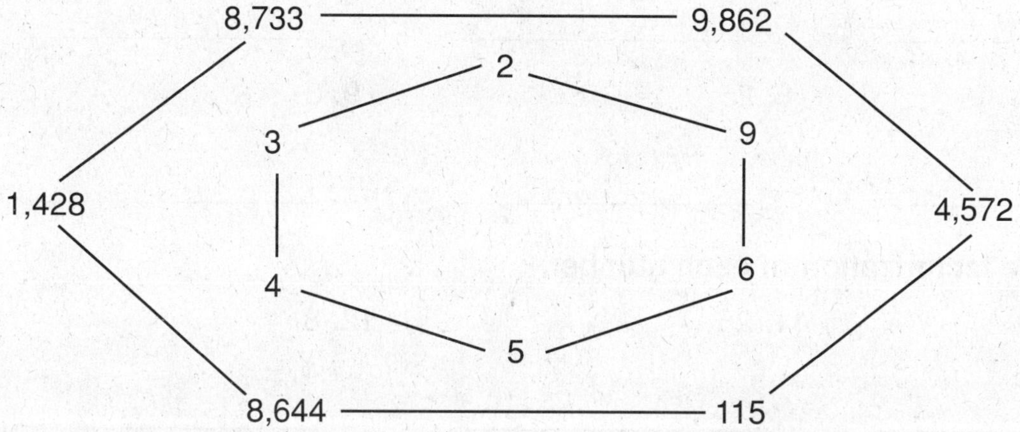

9,862 was divisible by ______ = P

8,733 was divisible by ______ = I

8,644 was divisible by ______ = S

4,572 was divisible by ______ = M

115 was divisible by ______ = T

To solve the riddle, fill in the letters that match your answers below.

What happened when Monkey 3 married Monkey 5?

They became

___	R	___	___	E	___	A	___	E	___
2		3	9		6		5		4

Name ______________________ Date __________ Class __________

LESSON 4-2

Practice A
Factors and Prime Factorization

List all of the factors of each number.

1. 6 ______

2. 9 ______

3. 10 ______

4. 12 ______

5. 21 ______

6. 18 ______

7. 16 ______

8. 25 ______

9. 31 ______

Write the prime factorization of each number.

10. 9 ______

11. 25 ______

12. 8 ______

13. 14 ______

14. 12 ______

15. 15 ______

16. 5 ______

17. 20 ______

18. 26 ______

19. There are 12 chairs in the meeting hall and an odd number of tables. What are all the possible ways the chairs could be arranged so that each table has the same number of chairs?

______

20. What are two different ways that 9 can be written as a product of two numbers?

______ ______

Name ______________________________ Date ____________ Class ____________

LESSON 4-2

Practice B

Factors and Prime Factorization

List all of the factors of each number.

1. 15

2. 24

3. 33

4. 72

5. 48

6. 95

7. 66

8. 87

9. 36

Write the prime factorization of each number.

10. 44

11. 56

12. 42

13. 39

14. 36

15. 125

16. 85

17. 100

18. 32

19. James has an assigned seat for his flight to Denver. The seats on the plane are numbered 1–49. James's seat number is an odd number greater than 10 that is factor of 100. What is his seat number for the flight?

20. Linda writes the prime factorization of 40 as $2 \cdot 2 \cdot 2 \cdot 5$ on the board. Phil writes the prime factorization of 40 as $2^3 \cdot 5$. Who is correct?

Name ______________________ Date __________ Class __________

LESSON **4-2**

Practice C

Factors and Prime Factorization

List all of the factors of each number.

1. 92

2. 356

3. 180

4. 550

Write the prime factorization of each number.

5. 225

6. 333

7. 124

8. 216

9. 423

10. 810

Write each number as a product in two different ways.

11. 81

12. 117

13. 375

14. You and your friend are going to split the houses in your neighborhood for newspaper delivery. Which would you prefer, the odd-numbered houses or the prime-numbered houses? Explain.

15. Selma cannot remember her locker number! She knows that her locker number is prime, and that it is a factor of 435. What are all of Selma's possible locker numbers?

Name ______________________ Date __________ Class __________

LESSON 4-2

Reteach

Factors and Prime Factorization

Factors of a product are the numbers that are multiplied to find that product. A factor is also a whole number that divides the product with no remainder.

To find all of the factors of 24, make a list of multiplication facts.

$1 \cdot 24 = 24$
$2 \cdot 12 = 24$
$3 \cdot 8 = 24$
$4 \cdot 6 = 24$

The factors of 24 are 1, 2, 3, 4, 6, 8, 12, and 24.

Write multiplication facts to find the factors of each number.

1. 20

____

____

2. 16

____

____

3. 35

____

____

4. 31

____

____

A number written as the product of prime factors is called the prime factorization of the number.

To write the prime factorization of 24, first write it as product of 2 numbers. Then rewrite each factor as the product of 2 numbers until all of the factors are prime numbers.

$24 = \mathbf{4} \cdot 6$ (Write 24 as the product of 2 numbers.)
$= 2 \cdot 2 \cdot \mathbf{6}$ (Rewrite 4 as the product of 2 prime numbers.)
$= 2 \cdot 2 \cdot 2 \cdot 3$ (Rewrite 6 as the product of 2 prime numbers.)

So, the prime factorization of 24 is $2 \cdot 2 \cdot 2 \cdot 3$ or $2^3 \cdot 3$.

Find the prime factorization of each number.

5. 28 __________

6. 45 __________

7. 50 __________

8. 72 __________

Name ______________________ Date __________ Class __________

LESSON 4-2

Challenge
Prime Shades

For each given number, shade one box in each row of the table to show a prime factor. Then use your shaded boxes to write each number's prime factorization.

1. 12

7	2
3	5
2	11

Prime factorization:

____

2. 70

5	3
2	11
3	7

Prime factorization:

____

3. 63

3	5
2	3
11	7

Prime factorization:

____

4. 150

13	5
3	7
2	11
17	5

Prime factorization:

____

5. 84

11	7
5	2
13	2
3	17

Prime factorization:

____

6. 260

5	11
3	2
17	13
2	7

Prime factorization:

____

7. 80

5	7
17	2
2	13
11	2
2	3

Prime factorization:

____

8. 1,750

17	5
13	7
31	5
5	3
2	11

Prime factorization:

____

9. 3,234

5	3
17	7
31	11
2	13
7	23

Prime factorization:

____

Name ______________________ Date __________ Class __________

LESSON 4-2

Problem Solving

Factors and Prime Factorization

Write the correct answer.

1. The area of a rectangle is the product of its length and width. If a rectangular board has an area of 30 square feet, what are the possible measurements of its length and width?

____

____

2. The first-floor apartments in Jenna's building are numbered 100 to 110. How many apartments on that floor are a prime number? What are those apartment numbers?

____

____

3. A Russian mathematician named Christian Goldbach came up with a theory that every even number greater than 4 can be written as the sum of two odd primes. Test Goldbach's theory with the numbers 6 and 50.

____

____

4. Mr. Samuels has 24 students in his math class. He wants to divide the students into equal groups, and he wants the number of students in each group to be prime. What are his choices for group sizes? How many groups can he make?

____

____

Circle the letter of the correct answer.

5. Why is 2 the only even prime number?

A It is the smallest prime number.
B All other even numbers are divisible by 2.
C It only has 1 and 2 as factors.
D All odd numbers are prime.

6. What prime numbers are factors of both 60 and 105?

F 2 and 3
G 2 and 5
H 3 and 5
J 5 and 7

7. If a composite number has the first five prime numbers as factors, what is the smallest number it could be? Write that number's prime factorization.

A 30
B 210
C 2,310
D 30,030

8. Tim's younger brother, Bryant, just had a birthday. Bryant's age only has one factor, and is not a prime number. How old is Bryant?

F 10 years old
G 7 years old
H 3 years old
J 1 year old

Name ______________________ Date ____________ Class ____________

LESSON **4-2**

Reading Strategies

Use an Organizer

A **prime number** has only two factors, 1 and the number itself.

2 ⟶ $2 \cdot 1 = 2$ The factors are 1 and 2.

A **composite number** has more than 2 factors.

6 ⟶ $1 \cdot 6 = 6$ and $2 \cdot 3 = 6$ The factors are 1, 2, 3, and 6.

When a composite number is written as the product of prime numbers, it is called a **prime factorization,** such as $12 = 2 \times 2 \times 3$.

Drawing a **factor tree** will help you find and organize the prime factors of a number such as 42.

42 ⟵ Write the number.

6 (7) ⟵ Write any pair of factors. Circle 7 because it is a prime number.

(2) (3) ⟵ Continue until all factors are prime.

$2 \times 3 \times 7$ ⟵ Write the prime factors from least to greatest.

Answer each question.

1. Write a pair of factors you could use to find the prime factorization of 30.

____

2. Use the factors in the answer above to write the prime factorization of 30.

____

3. Make a factor tree for 36. What pair of factors did you start with?

____

4. What is the prime factorization of 36?

____

5. Make a factor tree for 16. What pair of factors did you start with?

____

6. What is the prime factorization for 16?

____

Name ______________________ Date ____________ Class ____________

LESSON **4-2**

Puzzles, Twisters & Teasers

Get Cooking

For each number in the left-hand column, there is matching prime factorization in the right-hand column.

Connect the numbers on the left to the factorizations on the right for the correct match. Each line will pass through a letter in the center.

Fill in the letters below to solve the riddle.

1.	56		$3 \times 3 \times 5 \times 5$
		I	
2.	225	M	$2 \times 3 \times 7 \times 7$
		T	
3.	61	P	$2 \times 2 \times 2 \times 7$
		A	
4.	294	E	$3 \times 3 \times 3 \times 37$
		S	
5.	999	T	prime
		D	
6.	77	B	$2 \times 3 \times 3 \times 3 \times 5$
		R	
7.	41		7×11

What did the whole number serve at his family barbecue?

____ ____ ____ ____ ____
4 6 2 1 3

____ ____ ____ ____
6 2 7 5

Name ______________________ Date __________ Class __________

LESSON 4-3

Practice A
Greatest Common Factor

Find the GCF of each set of numbers.

1. 6 and 9 ________

2. 4 and 8 ________

3. 8 and 12 ________

4. 6 and 15 ________

5. 10 and 15 ________

6. 9 and 12 ________

7. 4 and 10 ________

8. 5 and 20 ________

9. 7 and 14 ________

10. 8 and 11 ________

11. 2 and 12 ________

12. 9 and 21 ________

Circle the letter of the correct answer.

13. For which set of numbers is 6 the GCF?

A 2, 3, and 6
B 3, 6, and 12
C 12, 18, and 24
D 1, 6, and 12

14. For which set of numbers is 4 the GCF?

F 1, 4, and 8
G 2, 4, and 16
H 1, 2, and 4
J 8, 12, and 16

15. Bonny has 24 wood beads and 30 glass beads. She wants each necklace she makes to have the same number of wood beads and the same number of glass beads. What is the greatest number of necklaces she can make if all the beads are used?

____

16. Mike is setting up fish tanks at the pet store. He has 6 angel fish, 12 tiger barbs, and 15 guppies. If he wants to have the same number of each kind of fish in every tank, what is the greatest number of tanks he can set up?

____

Name ______________________ Date ____________ Class ____________

LESSON **4-3**

Practice B

Greatest Common Factor

Find the GCF of each set of numbers.

1. 12 and 15 ____________

2. 18 and 24 ____________

3. 15 and 25 ____________

4. 16 and 24 ____________

5. 36 and 45 ____________

6. 24 and 54 ____________

7. 48 and 64 ____________

8. 27 and 72 ____________

9. 55 and 77 ____________

10. 16, 28, and 48 ____________

11. 15, 35, and 95 ____________

12. 20, 30, and 80 ____________

13. 18, 36, and 54 ____________

14. 27, 36, and 45 ____________

15. 21, 49, and 63 ____________

16. 25, 35, and 45 ____________

17. 28, 42, and 63 ____________

18. 25, 75, and 115 ____________

19. Mr. Thompson's sixth-grade class is competing in the school field day. There are 16 boys and 12 girls in his class. He divided the class into the greatest number of teams possible with the same number of boys on each team and the same number of girls on each team. How many teams were made if each person was on a team? How many girls were on each team? How many boys?

____

20. Barbara is making candy bags for her birthday party. She has 24 lollipops, 12 candy bars, and 42 pieces of gum. She wants each bag to have the same number of each kind of candy. What is the greatest number of bags she can make if all the candy is used? How many pieces of each kind of candy will be in each bag?

____

____

Name ______________________ Date ____________ Class ____________

LESSON 4-3

Practice C

Greatest Common Factor

Find the GCF of each set of numbers.

1. 48 and 64

2. 72 and 81

3. 54 and 66

4. 56 and 80

5. 36 and 48

6. 32 and 232

7. 20, 44, and 88

8. 72, 96, and 84

9. 54, 60, and 78

10. 49, 84, and 91

11. 150, 200, and 300

12. 88, 96, and 120

13. 81, 108, and 117

14. $3^3 \cdot 4$ and $3^2 \cdot 4^2$

15. $2^2 \cdot 3 \cdot 5$ and $2^3 \cdot 3^3$

16. 2, 8, 16, and 40

17. 12, 15, 20, and 30

18. 5, 10, 15, and 25

19. Mrs. Hill is preparing for group art projects. She has 52 pipe cleaners, 78 pieces of colored paper, 91 cotton balls, and 117 strings of yarn. Each group will get the same number of each kind of art supply. What is the greatest number of groups that Ms. Hill can make if all the supplies are used? How many of each item will each group get?

20. Tony is making flower arrangements. He is using three different kinds of flowers—roses, daisies, and tulips. Each arrangement will have the same number of each kind of flower. He figures that if he uses all the flowers, the greatest number of arrangements he can make is 7. He has 42 flowers in all, with more daisies than tulips, and more tulips than roses. How many of each kind of flower does Tony have?

Name ______________________________ Date ____________ Class ____________

LESSON 4-3

Reteach

Greatest Common Factor

The greatest common factor, or GCF, is the largest number that is the factor of any set of at least two numbers.

You can use prime factorization to find the GCF of two or more numbers.

To find the GCF of 18 and 24, first write the prime factorization of each number. Then identify the common prime factors.

18 = **2** • 3 • **3**
24 = **2** • 2 • 2 • **3**

Next, find the product of the common prime factors.
2 • 3 = 6

The GCF of 18 and 24 is 6.

Find the GCF of each set of numbers.

1. 32 and 48

32 = ____________

48 = ____________

____

2. 45 and 81

45 = ____________

81 = ____________

____

3. 18 and 36

18 = ____________

36 = ____________

____

4. 14 and 35

14 = ____________

35 = ____________

____

5. 42 and 72

42 = ____________

72 = ____________

____

6. 56 and 64

56 = ____________

64 = ____________

____

7. 28, 56, and 84

28 = ____________

56 = ____________

84 = ____________

____

8. 30, 45, and 75

30 = ____________

45 = ____________

75 = ____________

____

9. 36, 45, and 54

36 = ____________

45 = ____________

54 = ____________

____

Name ______________________________ Date ____________ Class ____________

LESSON 4-3

Challenge

The Greatest Common Flower

A florist made these flower arrangements for a wedding. He used every flower in each crate he had to create the greatest number of arrangements possible. Study the flowers the florist had in each crate on the left. Below each, write the number of arrangements the florist made with those flowers. Then draw a line to the correct arrangement on the right that the florist created with those flowers.

1. 

6 18 24

Number of arrangements: ______

2. 

12 16 20

Number of arrangements: ______

3. 

15 30 35

Number of arrangements: ______

4. 

21 42 56

Number of arrangements: ______

5. 

9 15 18

Number of arrangements: ______

A. 

B. 

C. 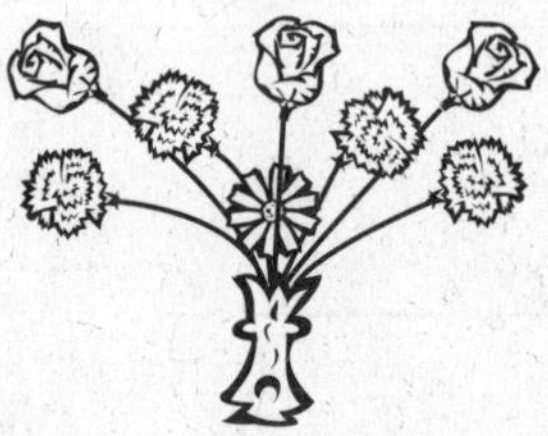

D. 

E. 

Name ______________________ Date ____________ Class ____________

LESSON 4-3

Problem Solving

Greatest Common Factor

Write the correct answer.

1. Carolyn has 24 bottles of shampoo, 36 tubes of hand lotion, and 60 bars of lavender soap to make gift baskets. She wants to have the same number of each item in every basket. What is the greatest number of baskets she can make without having any of the items left over?

____

____

2. There are 40 girls and 32 boys who want to participate in the relay race. If each team must have the same number of girls and boys, what is the greatest number of teams that can race? How many boys and girls will be on each team?

____

____

3. Ming has 15 quarters, 30 dimes, and 48 nickels. He wants to group his money so that each group has the same number of each coin. What is the greatest number of groups he can make? How many of each coin will be in each group? How much money will each group be worth?

____

____

____

4. A gardener has 27 tulip bulbs, 45 tomato plants, 108 rose bushes, and 126 herb seedlings to plant in the city garden. He wants each row of the garden to have the same number of each kind of plant. What is the greatest number of rows that the gardener can make if he uses all the plants?

____

____

____

Circle the letter of the correct answer.

5. Kim packed 6 boxes with identical supplies. It was the greatest number she could pack and use all the supplies. Which of these is her supply list?

A 24 pencils, 36 pens, 10 rulers

B 12 rulers, 30 pencils, 45 pens

C 42 pencils, 18 rulers, 72 pens

D 60 pens, 54 pencils, 32 rulers

6. The sum of three numbers is 60. Their greatest common factor is 4. Which of the following lists shows those three numbers?

F 4, 16, 36

G 8, 20, 32

H 14, 16, 30

J 10, 18, 32

Name ______________________ Date ____________ Class ____________

LESSON 4-3

Reading Strategies

Follow a Procedure

Numbers you multiply together are called **factors.** When you multiply factors together, the answer is called the **product.**

Answer each question.

1. Is 2 a factor of 8? Why or why not?

____

2. Is 3 a factor of 8? Why or why not?

____

The word **common** means "shared." All the students in a class who have blue eyes have eye color in common.
Whole numbers may share the same factors. Shared factors are called **common factors.**

These are the factor pairs for 12 and 18.

12:	1 • 12	2 • 6	3 • 4
18:	1 • 18	2 • 9	3 • 6

Answer the following questions.

3. List all of the factors for 12.

____

4. List all of the factors for 18.

____

5. Find the factors that 12 and 18 share in common.

____

The largest factor that two or more whole numbers share is called the **greatest common factor.**

6. What is the greatest common factor for 12 and 18?

____

7. List the steps you use to find the greatest common factor of two numbers.

____

____

Name ______________________________ Date ____________ Class ____________

LESSON **4-3**

Puzzles, Twisters & Teasers

Magical Factor Cruise

This is the game of Magical Factor Island. You are currently on a cruise ship and must get to this island in order to win a magical prize.

You must decide if the statement in each box is true or false. If it is false black out the box. Once you have all the false statements blacked out you will be able to cruise to the magical island.

GCF of 12 and 8 = 4	GCF of 5 and 12 = 5	GCF of 42 and 28 = 12	GCF of 63 and 18 = 18	GCF of 96 and 64 = 30
GCF of 6 and 24 = 6	GCF of 8 and 20 = 4	GCF of 12 and 48 = 12	GCF of 50 and 125 = 15	GCF of 70 and 20 = 5
GCF of 15 and 27 = 9	GCF of 12 and 72 = 4	GCF of 10 and 15 = 5	GCF of 48 and 36 = 4	GCF of 18 and 24 = 4
GCF of 32 and 48 = 2	GCF of 24 and 66 = 2	GCF of 24 and 32 = 8	GCF of 15 and 27 = 3	GCF of 6 and 18 = 6
GCF of 16 and 33 = 4	GCF of 6 and 18 = 3	GCF of 12 and 24 = 6	GCF of 24 and 56 = 2	GCF of 20 and 40 = 20
GCF of 9 and 45 = 5	GCF of 18 and 30 = 3	GCF of 75 and 90 = 10	GCF of 40 and 100 = 40	GCF of 16 and 40 = 8

Name ______________________ Date __________ Class __________

LESSON 4-4

Practice A

Decimals and Fractions

Write each decimal as a fraction or mixed number.

1. 0.5 ________

2. 0.25 ________

3. 0.75 ________

4. 0.4 ________

5. 0.8 ________

6. 1.2 ________

Write each fraction or mixed number as a decimal.

7. $\frac{3}{10}$ ________

8. $\frac{1}{2}$ ________

9. $\frac{1}{4}$ ________

10. $1\frac{3}{4}$ ________

11. $\frac{3}{5}$ ________

12. $1\frac{2}{5}$ ________

Circle the letter of the best answer.

13. Which of the following sets is written in order from least to greatest?

A 0.5, $\frac{1}{4}$, 0.75

B 0.4. $\frac{7}{10}$, 0.6

C $\frac{1}{4}$, 0.5, 0.75

D $\frac{7}{10}$, 0.4, 0.6

14. Which of the following sets is written in order from greatest to least?

F $\frac{1}{3}$, $1\frac{1}{2}$, $1\frac{3}{4}$

G $\frac{2}{5}$, $0.\overline{3}$, 0.3

H $1\frac{1}{2}$, $1\frac{3}{4}$, $1\frac{1}{3}$

J 0.3, $0.\overline{3}$, $\frac{2}{5}$

15. At Franklin Elementary School, $\frac{2}{3}$ of all the students attended the chorus recital on Thursday. On Friday, $\frac{3}{4}$ of all the students attended the basketball game. Which event had the highest attendance?

____

16. Each of the Girl Scout troops was given the same number of cookies to sell. Tina's troop sold 0.25 of its cookies. Chantall's Girl Scout troop sold half of its cookies. Which troop sold the most cookies?

____

Name ______________________________ Date ____________ Class ____________

LESSON 4-4

Practice B

Decimals and Fractions

Write each decimal as a fraction or mixed number.

1. 0.23 ____________

2. 0.1 ____________

3. 3.25 ____________

4. $1.\overline{3}$ ____________

5. 5.5 ____________

6. 3.7 ____________

Write each fraction or mixed number as a decimal.

7. $\frac{4}{5}$ ____________

8. $\frac{1}{9}$ ____________

9. $1\frac{2}{3}$ ____________

10. $3\frac{3}{5}$ ____________

11. $2\frac{1}{3}$ ____________

12. $\frac{8}{9}$ ____________

Order the fractions and decimals from least to greatest.

13. $\frac{1}{4}$, 0.7, $\frac{3}{5}$ ____________

14. 0.25, $\frac{1}{8}$, 0.3 ____________

15. $\frac{9}{10}$, 0.49, $\frac{1}{2}$ ____________

Order the fractions and decimals from greatest to least.

16. 0.13, $\frac{1}{10}$, 0.9 ____________

17. $\frac{2}{5}$, 0.7, $\frac{2}{3}$ ____________

18. 0.65, $\frac{4}{5}$, $\frac{3}{4}$ ____________

19. Derrick has a dollar bill and three dimes, Jane has a dollar bill and one quarter, and Kelly has a dollar bill and ten nickels. Who has the most money? the least?

____

20. It rained three and one half inches in April. In May it rained $3\frac{3}{4}$ inches, and in June it rained 3.6 inches. Write the months in order from the greatest to the least amount of rain.

____

Name Name Date Date Class Class

LESSON 4-4

Practice C

Decimals and Fractions

Write each decimal as a fraction or mixed number.

1. 0.97

2. 2.03

3. 56.1

4. $6.\overline{6}$

5. 7.53

6. 12.009

Write each fraction or mixed number as a decimal.

7. $1\frac{8}{9}$

8. $\frac{1}{12}$

9. $2\frac{7}{100}$

10. $\frac{31}{500}$

11. $\frac{17}{20}$

12. $\frac{5}{11}$

Order the fractions and decimals from least to greatest.

13. $0.83, \frac{7}{8}, \frac{4}{5}$

14. $\frac{9}{11}, 0.9, \frac{5}{6}$

15. $4\frac{2}{3}, 4.\overline{2}, 4\frac{3}{11}$

Order the fractions and decimals from greatest to least.

16. $\frac{3}{10}, \frac{27}{100}, \frac{1}{3}$

17. $\frac{8}{12}, \frac{3}{4}, 0.71$

18. $\frac{19}{20}, \frac{97}{100}, 0.99$

19. Shawn has 3.65 cups of sugar and 1.9 cups of flour. He needs $3\frac{2}{3}$ cups of sugar and $1\frac{7}{8}$ cups of flour to make cookies. Does he have enough of each ingredient?

20. This week, Katie ran $12\frac{11}{12}$ miles, Sandra ran 12.91 miles, and Tameeka ran $12\frac{9}{10}$ miles. Who ran the most miles this week? the least?

Name ______________________ Date ____________ Class ____________

LESSON 4-4

Reteach

Decimals and Fractions

You can write decimals as fractions or mixed numbers. A place value chart will help you read the decimal. Remember the decimal point is read as the word "and."

To write 0.47 as a fraction, first think about the decimal in words.

Ones	Tenths	Hundredths	Thousandths	Ten Thousandths
0.	4	7		

0.47 is read "forty-seven hundredths." The place value of the decimal tells you the denominator is 100.

$0.47 = \frac{47}{100}$

To write 8.3 as a mixed number, first think about the decimal in words.

Ones	Tenths	Hundredths	Thousandths	Ten Thousandths
8.	3			

8.3 is read "eight and three tenths." The place value of the decimal tells you the denominator is 10. The decimal point is read as the word "and."

$8.3 = 8\frac{3}{10}$

Write each decimal as a fraction or mixed number.

1. 0.61 ____________

2. 3.43 ____________

3. 0.009 ____________

4. 4.7 ____________

5. 1.5 ____________

6. 0.13 ____________

7. 5.002 ____________

8. 0.021 ____________

Name ______________________ Date __________ Class __________

LESSON 4-4

Reteach

Decimals and Fractions (continued)

Fractions and mixed numbers can be written as decimals.

To write $\frac{1}{4}$ as a decimal, first think about the expression in words.

$\frac{1}{4}$ means "1 divided by 4."

Then do the division.

$$\begin{array}{r} 0.25 \\ 4\overline{)1.00} \\ \underline{-8} \\ 20 \\ \underline{-\,20} \\ 0 \end{array}$$

0.25 is a terminating decimal because it has an end.

$\frac{1}{4} = 0.25$

A number that contains a whole number and a fraction is called a mixed number. $2\frac{1}{3}$ is an example of a mixed number. To write $2\frac{1}{3}$ as a decimal, first think about the expression in words. $2\frac{1}{3}$ means "2 and 1 divided by 3." Keep 2 as the whole number.

Then do the division.

$$\begin{array}{r} 0.33 \\ 3\overline{)1.00} \\ \underline{-9} \\ 10 \\ \underline{-09} \\ 1 \end{array}$$

$1 \div 3 = 0.33\ldots$, or $0.\overline{3}$.
$0.\overline{3}$ is a repeating decimal because it does not end.

So, $2\frac{1}{3} = 2.\overline{3}$.

Write each fraction or mixed number as a decimal.

9. $\frac{3}{5}$ ____________

10. $3\frac{3}{4}$ ____________

11. $\frac{2}{3}$ ____________

12. $1\frac{2}{9}$ ____________

13. $\frac{1}{6}$ ____________

14. $2\frac{1}{8}$ ____________

15. $\frac{5}{6}$ ____________

16. $8\frac{1}{9}$ ____________

Name ____________________ Date ____________ Class ____________

LESSON 4-4

Challenge

Fractions of Pizza

Write the fractions as decimals. Then slice and shade pieces of each pizza to represent each fraction. Finally, compare the pizzas in each row by writing <, >, or =.

1. $\frac{3}{4}$ ____________ ☐ $\frac{4}{5}$ ____________

2. $\frac{3}{6}$ ____________ ☐ $\frac{1}{2}$ ____________

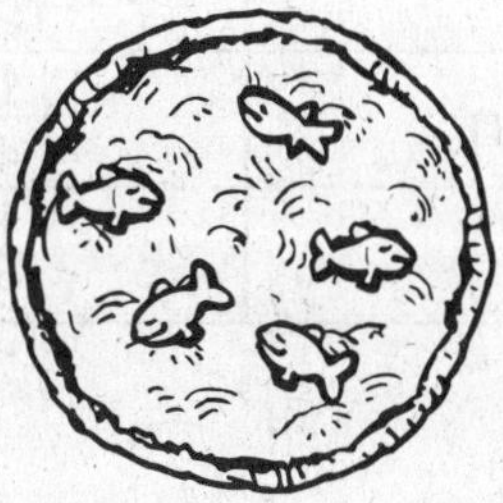

3. $\frac{2}{3}$ ____________ ☐ $\frac{3}{5}$ ____________

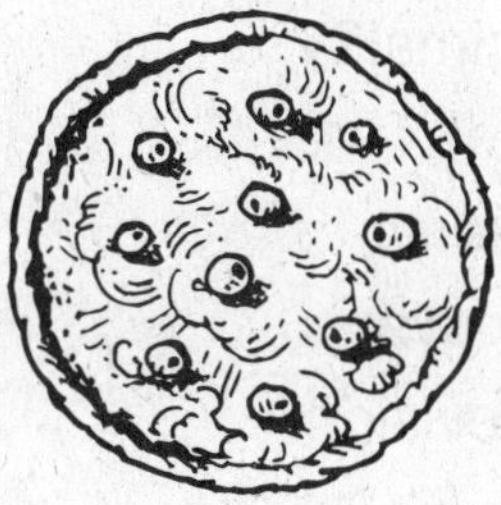
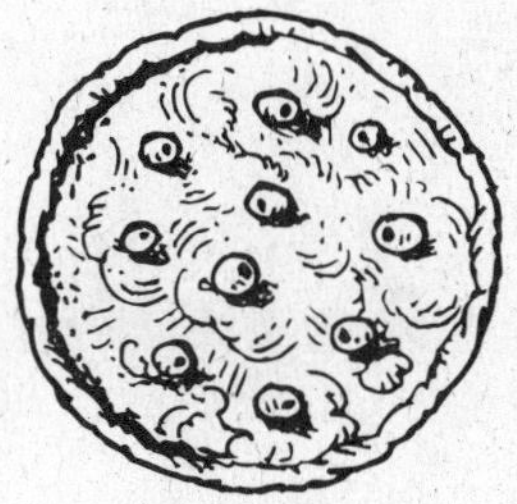

4. $\frac{1}{5}$ ____________ ☐ $\frac{1}{3}$ ____________

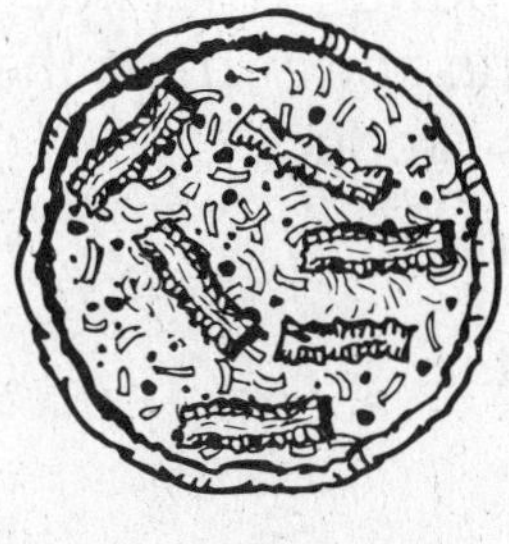
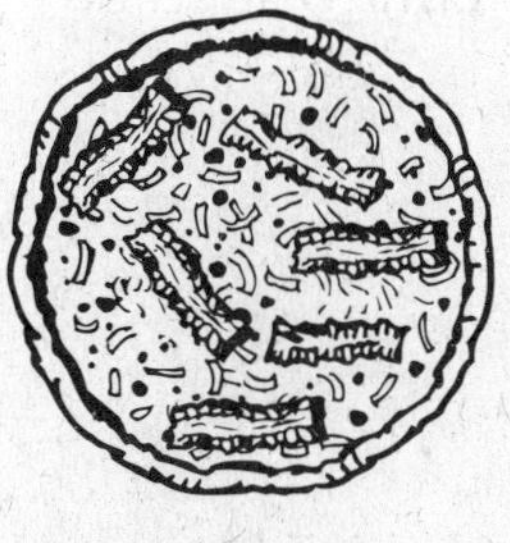

Name ______________________ Date ____________ Class ____________

LESSON 4-4

Problem Solving

Decimals and Fractions

Electricity is measured in amperes, or the rate electrical currents flow. A high ampere measurement means that a lot of electricity is being used. The table below shows the average amount of electricity some household appliances use per hour. Use the table to answer the questions.

Electricity Use in the Home

Appliance	Amps per Hour
Blender	$2\frac{1}{2}$
Coffeemaker	$6\frac{2}{3}$
Computer and printer	$1\frac{5}{6}$
Microwave oven	$12\frac{1}{2}$
Popcorn popper	$2\frac{1}{12}$
25-inch television	$1\frac{1}{4}$
VCR	$\frac{1}{3}$

1. How much electricity does an average 25-inch television use each hour? Write your answer as a decimal.

____

2. Which appliance uses an average of 2.5 amps per hour?

____

3. Which appliance uses the most electricity per hour? Write its ampere measurement as a decimal.

____

____

Circle the letter of the correct answer.

4. How much electricity do most computers and printers use in an hour?

   A 1.38 amperes
   B 1.8 amperes
   C $1.8\overline{3}$ amperes
   D 1.88 amperes

5. Which of the appliances has an hourly ampere measurement that is a repeating decimal?

   F blender
   G coffee maker
   H microwave oven
   J 25-inch television

6. In most years, 39.7 percent of the world's energy comes from burning oil. What is this percent written as a fraction?

   A $\frac{39}{7}$ percent
   B $39\frac{1}{7}$ percent
   C $3\frac{9}{7}$ percent
   D $39\frac{7}{10}$ percent

7. The United States produces about 13.2 percent of the world's hydroelectric power. What fraction of hydroelectric power does the United States produce?

   F $13\frac{1}{5}$ percent
   G $\frac{13}{2}$ percent
   H $1\frac{3}{2}$ percent
   J $13\frac{1}{2}$ percent

Name ______________________ Date ____________ Class ____________

LESSON 4-4

Reading Strategies

Focus on Vocabulary

The word *repeating* means something happening over and over. A church bell that chimes every hour is an example of a repeating sound.

The word *terminating* means something that ends. The sixth grade class party will be terminating at 10:00 P.M.

Fractions can be written as either **terminating** or **repeating decimals.**

In a repeating decimal, sometimes one digit repeats and sometimes more than one digit repeats.

$\frac{1}{3} \longrightarrow$ means $1 \div 3$

$$\begin{array}{r} 0.333 \\ 3\overline{)1.000} \\ -9 \\ \hline 10 \\ -9 \\ \hline 10 \\ -9 \\ \hline 1 \end{array}$$

$\frac{1}{3} = 0.\overline{3}$

The bar over 3 means the 3 keeps repeating.

$\frac{1}{4} \longrightarrow$ means $1 \div 4$

$$\begin{array}{r} 0.25 \\ 4\overline{)1.00} \\ -8 \\ \hline 20 \\ 20 \\ \hline 0 \end{array}$$

$\frac{1}{4} = 0.25$

0.25 is a decimal that terminates.

Write "terminating decimal" or "repeating decimal" for each. If the decimal repeats, write which digit or digits repeat.

1. 0.166666 ________________________________

2. 0.125 ________________________________

3. 0.090909 ________________________________

4. 0.2222 ________________________________

5. 0.8 ________________________________

6. 0.555 ________________________________

7. 0.5 ________________________________

8. 0.99 ________________________________

Name ______________________ Date __________ Class __________

LESSON 4-4

Puzzles, Twisters & Teasers

How High Can You Climb?

For each decimal, fraction and mixed number below there is a matching letter. Convert all of the numbers into a form so that you can order the numbers from smallest to largest. If you order the numbers correctly, the matching letters will spell out

2.123	T
$2\frac{1}{2}$	R
1.245	M
0.520	S
$\frac{6}{10}$	T
1.875	S
$\frac{9}{10}$	E
2.405	O
$1\frac{1}{4}$	O
$1\frac{1}{2}$	T
0.250	I
$2\frac{3}{4}$	E
$\frac{6}{8}$	H
0.340	T
$\frac{1}{2}$	A
2.645	I
$\frac{2}{5}$	H
$2\frac{4}{5}$	S
2.950	!
1.350	S

the answer to the riddle.

Why is the library the tallest room in the school?

Name ______________________ Date __________ Class __________

LESSON 4-5

Practice A

Equivalent Fractions

Find two equivalent fractions for each fraction.

1. $\frac{1}{2}$ ______ **2.** $\frac{1}{4}$ ______ **3.** $\frac{1}{3}$ ______

4. $\frac{1}{5}$ ______ **5.** $\frac{2}{3}$ ______ **6.** $\frac{3}{4}$ ______

7. $\frac{1}{6}$ ______ **8.** $\frac{2}{5}$ ______ **9.** $\frac{10}{20}$ ______

Find the missing numbers that make the fractions equivalent.

10. $\frac{2}{4} = \frac{?}{8}$ ______ **11.** $\frac{1}{3} = \frac{?}{9}$ ______ **12.** $\frac{10}{5} = \frac{?}{1}$ ______

13. $\frac{14}{4} = \frac{?}{2}$ ______ **14.** $\frac{5}{6} = \frac{?}{12}$ ______ **15.** $\frac{2}{7} = \frac{?}{21}$ ______

Write each fraction in simplest form.

16. $\frac{2}{4}$ ______ **17.** $\frac{3}{9}$ ______ **18.** $\frac{8}{10}$ ______ **19.** $\frac{4}{6}$ ______

20. A recipe calls for $\frac{1}{2}$ cup of sugar and $\frac{4}{8}$ cup of flour. Does the recipe use more sugar or flour? Explain.

____

21. Frank and Antoine's teams played the same number of baseball games this season. Frank's team won $\frac{2}{3}$ of its games, and Antoine's team won $\frac{8}{12}$ of its games. Which team won the most games this season? Explain.

____

Name ______________________ Date ____________ Class ____________

LESSON 4-5

Practice B

Equivalent Fractions

Find two equivalent fractions for each fraction.

1. $\frac{3}{6}$ ____________

2. $\frac{4}{7}$ ____________

3. $\frac{11}{13}$ ____________

4. $\frac{2}{15}$ ____________

5. $\frac{5}{14}$ ____________

6. $\frac{8}{9}$ ____________

7. $\frac{2}{21}$ ____________

8. $\frac{24}{48}$ ____________

9. $\frac{25}{100}$ ____________

Find the missing numbers that make the fractions equivalent.

10. $\frac{4}{7} = \frac{?}{28}$ ____________

11. $\frac{2}{9} = \frac{?}{54}$ ____________

12. $\frac{36}{4} = \frac{?}{1}$ ____________

13. $\frac{56}{8} = \frac{?}{2}$ ____________

14. $1\frac{3}{5} = \frac{?}{25}$ ____________

15. $1\frac{4}{7} = \frac{?}{42}$ ____________

Write each fraction in simplest form.

16. $\frac{15}{25}$ ____________

17. $\frac{8}{36}$ ____________

18. $\frac{12}{18}$ ____________

19. $\frac{10}{24}$ ____________

20. Billy had 24 trading cards. He gave 7 of his cards to Miko and 9 of his cards to Teri. What fraction of his original 24 cards does Billy have left? Write two equivalent fractions for that amount.

____

21. Beth and Kristine ride their bikes to school in the morning. Beth has to ride $1\frac{7}{32}$ miles. Kristine has to ride $\frac{39}{32}$ miles. Who rides the farthest to reach school? Explain.

____

Name ______________________ Date __________ Class __________

LESSON 4-5

Practice C

Equivalent Fractions

Find two equivalent fractions for each fraction.

1. $\frac{10}{15}$ ______
2. $\frac{7}{18}$ ______
3. $\frac{5}{16}$ ______
4. $\frac{9}{23}$ ______
5. $\frac{14}{27}$ ______
6. $\frac{155}{255}$ ______
7. $\frac{62}{80}$ ______
8. $\frac{21}{77}$ ______
9. $\frac{94}{102}$ ______

Find the missing numbers that make the fractions equivalent.

10. $\frac{2}{21} = \frac{?}{42}$ ______
11. $\frac{28}{56} = \frac{?}{14}$ ______
12. $\frac{15}{13} = \frac{?}{39}$ ______
13. $1\frac{10}{12} = \frac{?}{6}$ ______
14. $2\frac{3}{8} = \frac{?}{16}$ ______
15. $2\frac{4}{6} = \frac{?}{3}$ ______

Write each fraction in simplest form. Show two ways to simplify.

16. $\frac{18}{30}$ ______
17. $\frac{16}{24}$ ______
18. $\frac{16}{48}$ ______
19. $\frac{12}{39}$ ______

20. A theater manager has the same number of tickets available for every show. She sold $\frac{27}{45}$ of the tickets for Monday night's show. For Tuesday night's show, she sold $\frac{108}{180}$ of the tickets, and for Wednesday she sold $\frac{79}{135}$ of the tickets. For which two shows did she sell the same number of tickets? Explain.

____

21. What do the fractions $\frac{3}{17}$, $\frac{5}{7}$, $\frac{11}{23}$, and $\frac{83}{97}$ have in common? Why is it impossible to write those fractions in simplest form?

____

____

Name ______________________ Date ____________ Class ____________

LESSON **4-5**

Reteach

Equivalent Fractions

Fractions that have the same value are equivalent fractions.

$\frac{3}{4} = \frac{6}{8}$

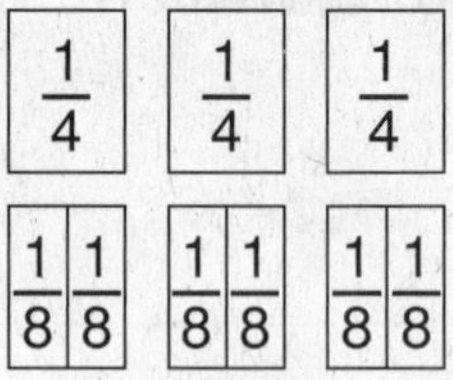

$\frac{3}{4}$ and $\frac{6}{8}$ are equivalent fractions.

You can use fraction strips to help you find equivalent fractions.

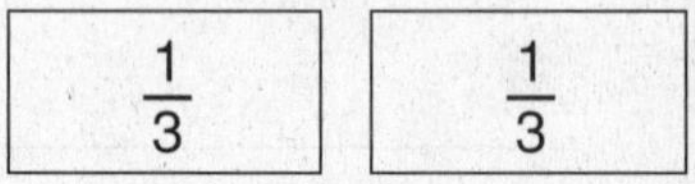

To solve $\frac{2}{3} = \frac{?}{12}$, first use fraction strips to model the first fraction.

Then use $\frac{1}{12}$ fraction pieces to make a length as long as the $\frac{2}{3}$ strip.

You need eight $\frac{1}{12}$ pieces, so $\frac{2}{3} = \frac{8}{12}$.

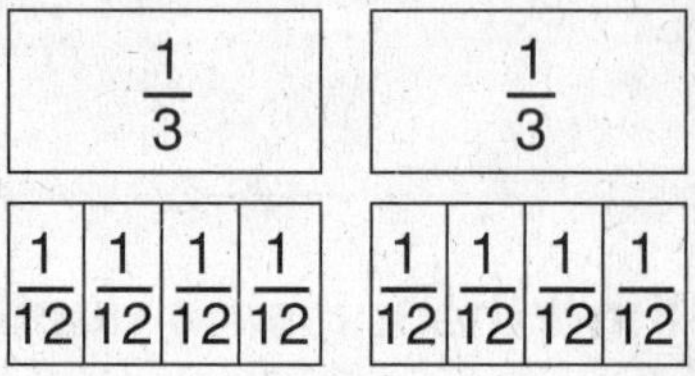

Find the missing number that makes the fractions equivalent.

1. $\frac{3}{4} = \frac{?}{12}$ ____________ **2.** $\frac{8}{10} = \frac{?}{5}$ ____________ **3.** $\frac{6}{9} = \frac{?}{3}$ ____________ **4.** $\frac{5}{6} = \frac{?}{12}$ ____________

A fraction is in simplest form when the GCF of the numerator and the denominator is 1.

To write $\frac{4}{6}$ in simplest form, divide the numerator and the denominator by their GCF, 2.

$4 \div 2 = 2$

$6 \div 2 = 3$

So, $\frac{4}{6}$ in simplest form is $\frac{2}{3}$.

Write each fraction in simplest form.

5. $\frac{3}{9}$ ____________ **6.** $\frac{12}{16}$ ____________ **7.** $\frac{14}{18}$ ____________ **8.** $\frac{8}{20}$ ____________

Name ______________________ Date __________ Class __________

LESSON 4-5

Challenge

Match That Fraction!

Write the fraction that the shaded part of each figure represents. Then match equivalent fractions by drawing lines between numbered and lettered figures.

1.

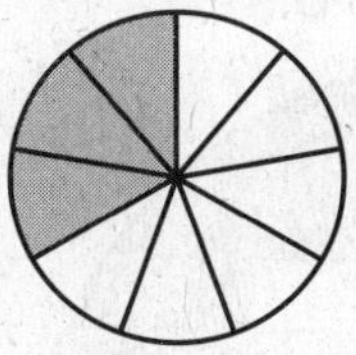

Fraction: __________

A.

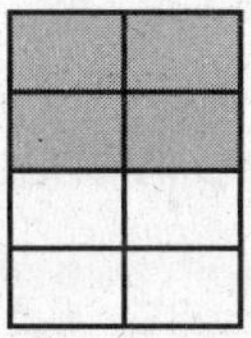

Fraction: __________

2.

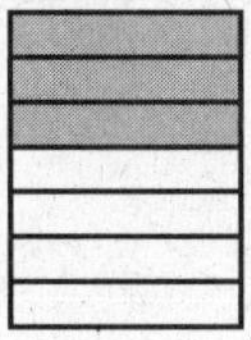

Fraction: __________

B.

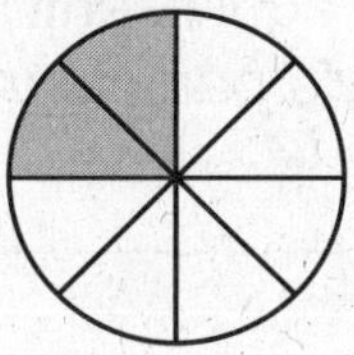

Fraction: __________

3.

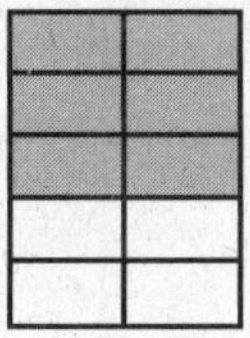

Fraction: __________

C.

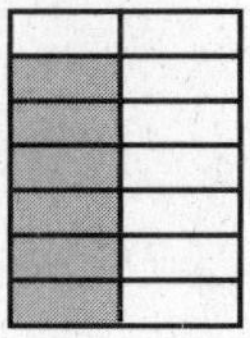

Fraction: __________

4.

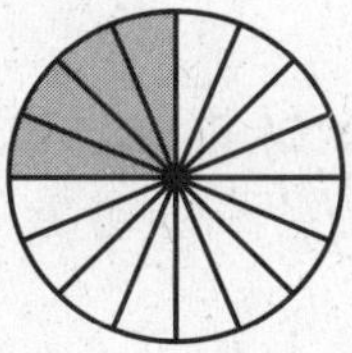

Fraction: __________

D.

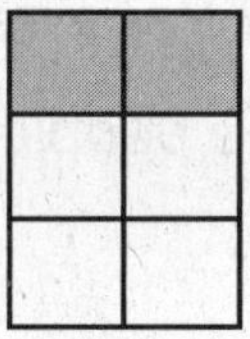

Fraction: __________

5.

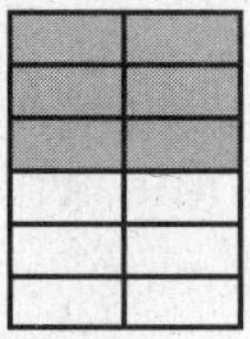

Fraction: __________

E.

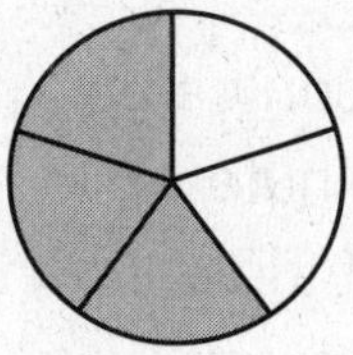

Fraction: __________

Name ______________________ Date ____________ Class ____________

LESSON 4-5

Problem Solving

Equivalent Fractions

About 60 million Americans exercise 100 times or more each year. Their top activities and the fraction of those 60 million people who did them are shown on the circle graph. Use the graph to answer the questions.

Exercise in the U.S.

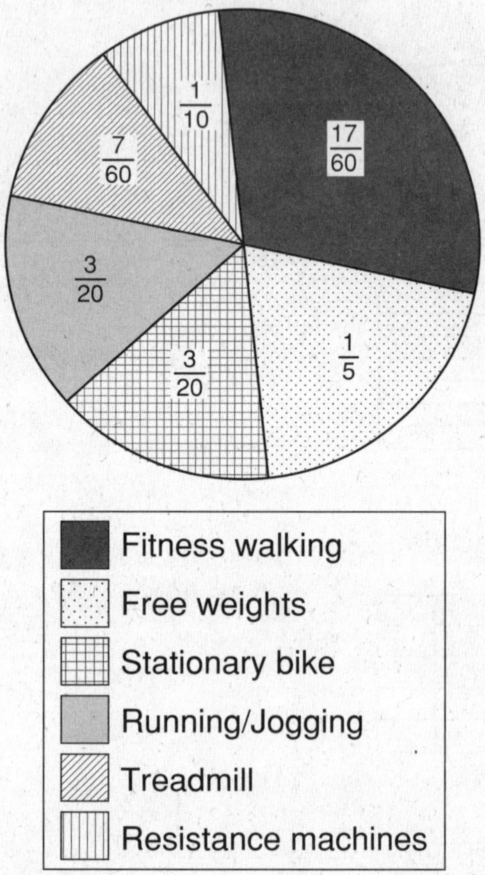

1. Which two activities did the same number of people use to keep in shape?

____

____

2. Which activity had the most participants? Write an equivalent fraction for that activity's participants.

____

____

3. Which activity had the fewest participants? Write two equivalent fractions for that activity's participants.

____

____

Circle the letter of the correct answer.

4. Which activity did $\frac{3}{15}$ of the people use to exercise?

A free weights

B treadmill

C fitness walking

D stationary bike

5. Which activity did $\frac{35}{300}$ of the people use to stay healthy?

F running/jogging

G resistance machines

H free weights

J treadmill

6. An average-sized person can burn about $6\frac{1}{2}$ calories a minute while riding a bike. Which of the following is equivalent to that amount?

A $1\frac{2}{2}$ **C** $6\frac{2}{4}$

B $5\frac{6}{2}$ **D** $6\frac{2}{6}$

7. An average-sized person can burn about 11.25 calories a minute while jogging. Which of the following is not equivalent to that amount?

F $11\frac{1}{4}$ **H** $11\frac{2}{8}$

G $11\frac{1}{2}$ **J** $11\frac{3}{12}$

Name ______________________ Date ____________ Class ____________

LESSON 4-5

Reading Strategies

Use a Visual Model

Each fraction has the same amount shaded. Fractions of equal value are called **equivalent fractions.**

A B C

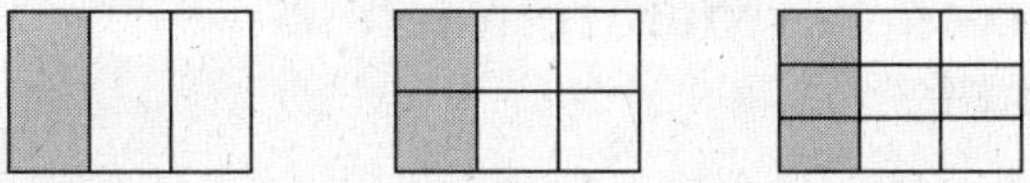

$$\frac{1}{3} = \frac{2}{6} = \frac{3}{9}$$

Answer the following questions.

1. What fractional part of rectangle A is shaded? ____________

2. What fractional part of rectangle B is shaded? ____________

3. What fractional part of rectangle C is shaded? ____________

4. Write two equivalent fractions for $\frac{1}{3}$. ____________

D E F

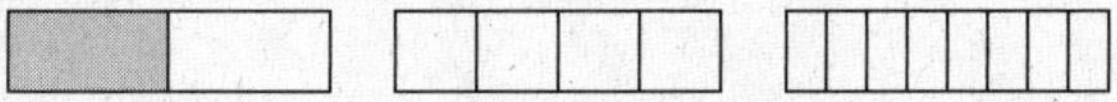

5. What fractional part of rectangle D is shaded? ____________

6. Shade an equivalent fractional part of rectangle E.

7. Shade an equivalent fractional part of rectangle F.

8. Write two equivalent fractions for $\frac{1}{2}$. ____________

Name ______________________ Date ____________ Class ____________

LESSON **4-5**

Puzzles, Twisters & Teasers

The Best Day in School

Your goal is to find all of the fractions in the grid which are equivalent.

Black out any pair or set of three fractions that are equivalent. Read the letters in the remaining boxes in order from left to right, top to bottom, to solve the riddle.

$\frac{6}{10}$ **W**	$\frac{12}{32}$ **C**	$\frac{45}{55}$ **D**	$\frac{7}{14}$ **A**	$\frac{18}{45}$ **O**	$\frac{21}{27}$ **L**
$\frac{18}{48}$ **E**	$\frac{2}{14}$ **E**	$\frac{2}{16}$ **V**	$\frac{6}{21}$ **X**	$\frac{36}{40}$ **M**	$\frac{12}{26}$ **F**
$\frac{20}{25}$ **G**	$\frac{15}{18}$ **T**	$\frac{9}{27}$ **L**	$\frac{5}{40}$ **S**	$\frac{2}{18}$ **I**	$\frac{15}{25}$ **T**
$\frac{21}{36}$ **T**	$\frac{16}{20}$ **H**	$\frac{46}{92}$ **S**	$\frac{35}{45}$ **P**	$\frac{18}{39}$ **N**	$\frac{21}{27}$ **K**
$\frac{14}{24}$ **Y**	$\frac{4}{44}$ **G**	$\frac{8}{18}$ **H**	$\frac{6}{24}$ **E**	$\frac{8}{28}$ **B**	$\frac{24}{36}$ **T**
$\frac{14}{20}$ **E**	$\frac{6}{24}$ **O**	$\frac{4}{10}$ **U**	$\frac{24}{28}$ **D**	$\frac{55}{66}$ **I**	$\frac{90}{100}$ **E**

What happened when the power went off at school?

The students were ______________!

Name ______________________ Date __________ Class __________

LESSON 4-6

Practice A

Mixed Numbers and Improper Fractions

Determine whether each fraction is proper or improper.

1. $\frac{1}{4}$ ________

2. $\frac{3}{4}$ ________

3. $\frac{7}{4}$ ________

4. $\frac{9}{3}$ ________

5. $\frac{2}{3}$ ________

6. $\frac{6}{2}$ ________

Write each mixed number as an improper fraction.

7. $1\frac{1}{2}$ ________

8. $1\frac{1}{3}$ ________

9. $2\frac{1}{2}$ ________

10. $1\frac{1}{4}$ ________

11. $1\frac{2}{3}$ ________

12. $2\frac{1}{3}$ ________

Write each improper fraction as a mixed number or whole number. Tell whether your answer is a mixed number or whole number.

13. $\frac{4}{1}$ ________

14. $\frac{6}{3}$ ________

15. $\frac{4}{3}$ ________

16. $\frac{5}{2}$ ________

17. $\frac{7}{5}$ ________

18. $\frac{3}{2}$ ________

19. Mickey timed her speech at $\frac{12}{5}$ minutes. She cannot go over 2 minutes. Should she add more to her speech, or cut some lines? Explain.

____

____

20. In a frog-jumping contest, Jim's frog hopped $2\frac{1}{3}$ feet. Matthew's frog hopped $1\frac{2}{3}$ feet. Whose frog hopped $\frac{7}{3}$ feet?

____

Name ______________________ Date ____________ Class ____________

LESSON 4-6

Practice B

Mixed Numbers and Improper Fractions

Write each mixed number as an improper fraction.

1. $3\frac{1}{2}$ ____________

2. $2\frac{1}{3}$ ____________

3. $5\frac{1}{4}$ ____________

4. $1\frac{3}{7}$ ____________

5. $3\frac{3}{4}$ ____________

6. $4\frac{1}{3}$ ____________

7. $2\frac{3}{5}$ ____________

8. $3\frac{5}{6}$ ____________

9. $7\frac{1}{3}$ ____________

Write each improper fraction as a mixed number or whole number. Tell whether your answer is a mixed number or whole number.

10. $\frac{17}{3}$ ____________

11. $\frac{40}{8}$ ____________

12. $\frac{48}{7}$ ____________

13. $\frac{33}{10}$ ____________

14. $\frac{50}{8}$ ____________

15. $\frac{83}{9}$ ____________

16. $\frac{104}{8}$ ____________

17. $\frac{121}{6}$ ____________

18. $\frac{78}{11}$ ____________

19. The hotel ordered an extra-long rug for a hallway that is $\frac{123}{2}$ feet long. What is the rug's length in feet and inches? Remember, 1 foot = 12 inches.

____

20. During this year's football-throwing contest, John threw the ball $49\frac{2}{3}$ feet. Sharon threw the ball 51 feet. Who threw the ball $\frac{153}{3}$ feet?

____

Name ______________________ Date __________ Class __________

LESSON 4-6

Practice C

Mixed Numbers and Improper Fractions

Write each mixed number as an improper fraction.

1. $2\frac{2}{5}$ ________

2. $1\frac{5}{10}$ ________

3. $6\frac{3}{4}$ ________

4. $2\frac{5}{9}$ ________

5. $14\frac{2}{3}$ ________

6. $2\frac{9}{11}$ ________

Write each improper fraction as a mixed number or whole number. Tell whether your answer is a mixed number or whole number.

7. $\frac{161}{7}$ ________

8. $\frac{73}{4}$ ________

9. $\frac{52}{9}$ ________

10. $\frac{225}{9}$ ________

11. $\frac{234}{11}$ ________

12. $\frac{127}{12}$ ________

Replace each shape with a number that will make the equation correct.

13. $\blacktriangle\frac{2}{7} = \frac{93}{\blacklozenge}$ ________

14. $\blacktriangle\frac{3}{8} = \frac{67}{\blacklozenge}$ ________

15. $\blacktriangle\frac{4}{5} = \frac{64}{\blacklozenge}$ ________

16. $43\frac{\blacktriangle}{15} = \frac{647}{\blacklozenge}$ ________

17. $13\frac{\blacktriangle}{9} = \frac{124}{\blacklozenge}$ ________

18. $14\frac{\blacktriangle}{8} = \frac{113}{\blacklozenge}$ ________

19. Amy made a video that is 5,400 seconds long. What is the video's running time in hours and minutes?

____

20. Justin accidentally erased the numerator of the fraction part of a mixed number. He knows that its whole number part is 45 and the denominator of the fraction part is 9. He also knows that when the mixed number is written as an improper fraction, its numerator is 412. What is the numerator that Justin erased?

____

Name ______________________ Date __________ Class __________

LESSON 4-6

Reteach

Mixed Numbers and Improper Fractions

A proper fraction is a fraction whose numerator is less than its denominator.

$\frac{2}{3}$, $\frac{1}{4}$, and $\frac{2}{7}$ are examples of proper fractions.

An improper fraction is a fraction whose numerator is greater than or equal to its denominator.

$\frac{3}{2}$, $\frac{8}{3}$, and $\frac{5}{5}$ are examples of improper fractions.

Some improper fractions can be written as mixed numbers.

To write $\frac{7}{4}$ as a mixed number, draw circles divided into $\frac{1}{4}$ sections.

Then shade in 7 of the $\frac{1}{4}$ sections.

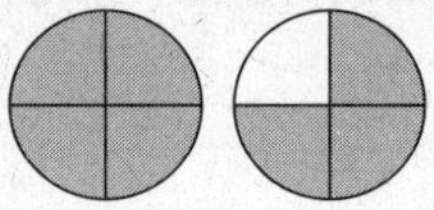

There is one circle and $\frac{3}{4}$ of a circle shaded.

So, $\frac{7}{4} = 1\frac{3}{4}$.

Write each improper fraction as a mixed number.

1. $\frac{14}{3}$ ________
2. $\frac{11}{2}$ ________
3. $\frac{15}{4}$ ________
4. $\frac{19}{6}$ ________

Mixed numbers can be written as improper fractions.

To write $2\frac{1}{3}$ as an improper fraction, draw 3 circles. Divide each circle into $\frac{1}{3}$ sections. Next, shade in 2 whole circles and one $\frac{1}{3}$ section of the last circle.

Then find the total number of $\frac{1}{3}$ sections that are shaded.

Seven $\frac{1}{3}$ sections are shaded, so $2\frac{1}{3} = \frac{7}{3}$.

Write each mixed number as an improper fraction.

5. $3\frac{1}{4}$ ________
6. $5\frac{2}{3}$ ________
7. $4\frac{1}{2}$ ________
8. $1\frac{5}{6}$ ________

Name ______________________________ Date ____________ Class ____________

LESSON **4-6**

Challenge

Mixed Up Mixed Numbers

Find the whole number and fraction in the box to make a mixed number to match each improper fraction listed below. Cross out each whole number and fraction as you use it. When you have finished, the unused items in the box will form the mixed number that answers the question at the bottom of the page.

1. $\frac{7}{3}$ = ____________ 2. $\frac{20}{3}$ = ____________

3. $\frac{21}{4}$ = ____________ 4. $\frac{31}{4}$ = ____________

5. $\frac{41}{5}$ = ____________ 6. $\frac{57}{5}$ = ____________

7. $\frac{59}{6}$ = ____________ 8. $\frac{19}{6}$ = ____________

9. $\frac{13}{7}$ = ____________ 10. $\frac{72}{7}$ = ____________

How old is the moon? ____________ billion years old!

Name ______________________ Date ____________ Class ____________

LESSON 4-6

Problem Solving

Mixed Numbers and Improper Fractions

Write the correct answer.

1. If stretched end-to-end, the total length of the blood vessels inside your body could wrap around Earth's equator $\frac{5}{2}$ times! Write this fact as a mixed number.

____

2. In 2000, the average 12-year-old child in the United States earned an allowance of 9 dollars and $\frac{7}{25}$ cents a week. Write this amount as an improper fraction and a decimal.

____

3. The normal body temperature for a rattlesnake is between $53\frac{3}{5}$°F and $64\frac{2}{5}$°F. Write this range as improper fractions.

____

4. A professional baseball can weigh no less than $\frac{45}{9}$ ounces and no more than $\frac{21}{4}$ ounces. Write this range as mixed numbers.

____

Circle the letter of the correct answer.

5. Betty needs a piece of lumber that is $\frac{14}{3}$ feet long. Which size should she look for at the hardware store?

A $3\frac{1}{3}$ feet

B $3\frac{1}{4}$ feet

C $4\frac{2}{3}$ feet

D $4\frac{1}{4}$ feet

6. What operations are used to change a mixed number to an improper fraction?

F multiplication and addition

G division and subtraction

H division and addition

J multiplication and subtraction

7. Adult bees only eat nectar, the substance in flowers used to make honey. A bee could fly 4 million miles on the energy it would get from eating $\frac{9}{2}$ liters of nectar. What is this amount of nectar written as a mixed number.

A $9\frac{1}{2}$ liters

B $4\frac{1}{2}$ liters

C $4\frac{1}{9}$ liters

D $2\frac{1}{2}$ liters

8. An astronaut who weighs 250 pounds on Earth would weigh $41\frac{1}{2}$ pounds on the moon. What is the astronaut's moon weight written as an improper fraction?

F $\frac{41}{2}$ pounds

G $\frac{42}{2}$ pounds

H $\frac{82}{2}$ pounds

J $\frac{83}{2}$ pounds

Name ______________________ Date ____________ Class ____________

LESSON 4-6

Reading Strategies

Focus on Vocabulary

A **mixed number** is made up of a whole number and a fraction.

Read: ⟶ two and one-half

Write: ⟶ $2\frac{1}{2}$

In an **improper fraction,** the numerator is greater than or equal to the denominator.

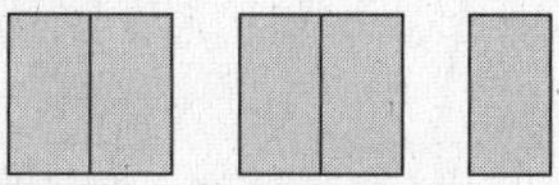

Read: ⟶ five halves

Write: ⟶ $\frac{5}{2}$

Answer the following questions.

1. What is a mixed number?

____

2. Why do you think it is called a "mixed" number?

____

____

3. What is an improper fraction?

____

4. Why do you think it is called an "improper" fraction?

____

Write "mixed number" or "improper fraction" for each number.

5. $1\frac{4}{5}$

____

6. $\frac{15}{8}$

____

7. $\frac{20}{17}$

____

8. $3\frac{2}{3}$

____

Name ______________________ Date __________ Class __________

LESSON 4-6

Puzzles, Twisters & Teasers

Surfing the Web

When you send information over the Internet it does not go in a straight path. Get your information from your computer to your friend by finding out the path it will take from Send to Receive.

You must always move from a smaller number to a larger number. You may move left, right, up and down only (no diagonal moves).

Send

$\frac{1}{8}$	$\frac{1}{16}$	$\frac{7}{3}$	$2\frac{7}{8}$	$\frac{17}{5}$	$2\frac{15}{16}$
$\frac{3}{16}$	$1\frac{1}{2}$	$1\frac{3}{5}$	$1\frac{1}{8}$	$3\frac{3}{4}$	$\frac{25}{8}$
$\frac{1}{8}$	$\frac{5}{4}$	$1\frac{1}{2}$	$\frac{16}{4}$	$\frac{33}{8}$	$4\frac{1}{16}$
$\frac{25}{5}$	$\frac{47}{9}$	$4\frac{5}{6}$	$\frac{9}{2}$	$4\frac{3}{16}$	$\frac{31}{8}$
$5\frac{1}{6}$	$5\frac{1}{3}$	$\frac{9}{2}$	$4\frac{1}{3}$	$4\frac{1}{8}$	$4\frac{8}{9}$
$\frac{11}{2}$	$\frac{35}{6}$	$\frac{25}{4}$	$6\frac{2}{7}$	$6\frac{1}{4}$	$\frac{27}{5}$
$\frac{20}{3}$	$5\frac{5}{9}$	$\frac{18}{3}$	$\frac{14}{2}$	$6\frac{1}{2}$	$6\frac{7}{11}$
$\frac{31}{4}$	$\frac{58}{7}$	$7\frac{6}{7}$	$8\frac{5}{8}$	$8\frac{7}{16}$	$\frac{23}{3}$
$8\frac{1}{8}$	$\frac{17}{2}$	$\frac{88}{11}$	$\frac{37}{4}$	$\frac{9}{1}$	$\frac{70}{7}$
$9\frac{1}{16}$	$9\frac{1}{8}$	$9\frac{3}{16}$	$9\frac{2}{5}$	$\frac{59}{6}$	$10\frac{1}{5}$

Receive

Name ______________________ Date ____________ Class ____________

LESSON 4-7

Practice A

Comparing and Ordering Fractions

Compare. Write <, >, or =.

1. $\frac{1}{6}$ ☐ $\frac{2}{6}$

2. $\frac{1}{3}$ ☐ $\frac{2}{3}$

3. $\frac{2}{4}$ ☐ $\frac{1}{2}$

4. $\frac{1}{2}$ ☐ $\frac{3}{4}$

5. $\frac{4}{6}$ ☐ $\frac{2}{3}$

6. $\frac{1}{5}$ ☐ $\frac{2}{3}$

7. $\frac{1}{2}$ ☐ $\frac{2}{5}$

8. $\frac{3}{4}$ ☐ $\frac{3}{5}$

9. $\frac{1}{9}$ ☐ $\frac{1}{8}$

Order the fractions from least to greatest.

10. $\frac{7}{9}, \frac{1}{9}, \frac{2}{9}$

____

11. $\frac{1}{3}, \frac{1}{4}, \frac{2}{3}$

____

12. $\frac{2}{5}, \frac{1}{2}, \frac{1}{10}$

____

Order the fractions from greatest to least.

13. $\frac{1}{5}, \frac{3}{5}, \frac{2}{5}$

____

14. $\frac{2}{7}, \frac{6}{7}, \frac{3}{6}$

____

15. $\frac{1}{6}, \frac{2}{3}, \frac{5}{6}$

____

16. The Mountain Cats and the Pirates have both played the same number of games. The Mountain Cats have won $\frac{1}{5}$ of its games, and the Pirates have won $\frac{1}{4}$ of its games. Which team has the most wins?

____

17. Each week, Kelly saves $\frac{1}{2}$ of her allowance. She spends $\frac{1}{3}$ of her allowance on food, and $\frac{1}{6}$ of it on games and movies. What does Kelly do with the largest part of her allowance?

____

Name ______________________ Date ____________ Class ____________

LESSON 4-7

Practice B

Comparing and Ordering Fractions

Compare. Write <, >, or =.

1. $\frac{4}{7}$ ☐ $\frac{3}{5}$

2. $\frac{1}{8}$ ☐ $\frac{2}{3}$

3. $\frac{1}{4}$ ☐ $\frac{2}{5}$

4. $\frac{7}{8}$ ☐ $\frac{5}{6}$

5. $\frac{18}{24}$ ☐ $\frac{3}{4}$

6. $\frac{4}{5}$ ☐ $\frac{8}{12}$

Order the fractions from least to greatest.

7. $\frac{1}{2}, \frac{2}{5}, \frac{1}{3}$

____

8. $\frac{2}{5}, \frac{3}{4}, \frac{2}{3}$

____

9. $\frac{3}{7}, \frac{5}{6}, \frac{4}{5}$

____

10. $\frac{5}{9}, \frac{3}{7}, \frac{2}{3}$

____

11. $\frac{3}{8}, \frac{2}{7}, \frac{3}{5}$

____

12. $\frac{2}{7}, \frac{1}{8}, \frac{2}{5}$

____

Order the fractions from greatest to least.

13. $\frac{1}{6}, \frac{2}{7}, \frac{1}{5}$

____

14. $\frac{3}{7}, \frac{4}{9}, \frac{2}{3}$

____

15. $\frac{2}{5}, \frac{3}{10}, \frac{2}{3}$

____

16. $\frac{4}{5}, \frac{7}{10}, \frac{1}{12}$

____

17. $\frac{3}{8}, \frac{3}{4}, \frac{4}{9}$

____

18. $\frac{4}{7}, \frac{3}{5}, \frac{5}{6}$

____

19. David ran $4\frac{1}{4}$ miles, Shane ran $4\frac{1}{2}$ miles, and Matt ran $4\frac{5}{8}$ miles. Who ran the farthest?

____

20. Darius and Anita both took the same test. Darius answered $\frac{5}{6}$ of the questions correctly, and Anita answered $\frac{6}{7}$ correctly. Who got the higher score on the test?

____

Name ______________________ Date __________ Class __________

LESSON 4-7

Practice C

Comparing and Ordering Fractions

Compare. Write <, >, or =.

1. $\frac{4}{9}$ ☐ $\frac{5}{7}$

2. $\frac{5}{12}$ ☐ $\frac{13}{22}$

3. $\frac{13}{15}$ ☐ $\frac{6}{7}$

4. $\frac{16}{35}$ ☐ $\frac{12}{20}$

5. $\frac{13}{16}$ ☐ $\frac{39}{48}$

6. $\frac{9}{11}$ ☐ $\frac{7}{8}$

Order the fractions from least to greatest.

7. $\frac{3}{8}, \frac{2}{7}, \frac{4}{9}$ ______________

8. $\frac{6}{11}, \frac{4}{5}, \frac{3}{4}$ ______________

9. $\frac{2}{7}, \frac{3}{9}, \frac{6}{7}$ ______________

10. $\frac{9}{10}, \frac{5}{6}, \frac{2}{3}$ ______________

11. $\frac{13}{15}, \frac{8}{9}, \frac{5}{8}$ ______________

12. $\frac{6}{7}, \frac{7}{10}, \frac{12}{13}$ ______________

Order the fractions from greatest to least.

13. $\frac{2}{3}, \frac{5}{6}, \frac{4}{7}$ ______________

14. $\frac{7}{10}, \frac{5}{7}, \frac{11}{13}$ ______________

15. $\frac{9}{15}, \frac{5}{8}, \frac{10}{13}$ ______________

16. $\frac{11}{17}, \frac{5}{8}, \frac{7}{9}$ ______________

17. $\frac{5}{9}, \frac{6}{7}, \frac{5}{6}$ ______________

18. $\frac{2}{9}, \frac{3}{8}, \frac{4}{7}$ ______________

19. A group of students collected $1,724 for the Red Cross. Bob collected $\frac{7}{24}$ of the money, Grady collected $\frac{3}{8}$, and Vania collected $\frac{1}{3}$. Who collected the most money?

____

20. The Gators soccer team plays the same number of games each year. The team won $\frac{9}{11}$ of its games in 2000, $\frac{81}{99}$ of its games in 2001, and $\frac{63}{77}$ of its games in 2002. In which year did the team win the most games?

____

Name ______________________ Date ____________ Class ____________

LESSON 4-7

Reteach

Comparing and Ordering Fractions

$\frac{3}{4}$ and $\frac{2}{3}$ are unlike fractions because they have different denominators. To compare these fractions, graph the fractions on the same number line.

The fraction that is farther to the right on the number line is greater in value.

$\frac{9}{12} > \frac{8}{12}$, so $\frac{3}{4} > \frac{2}{3}$.

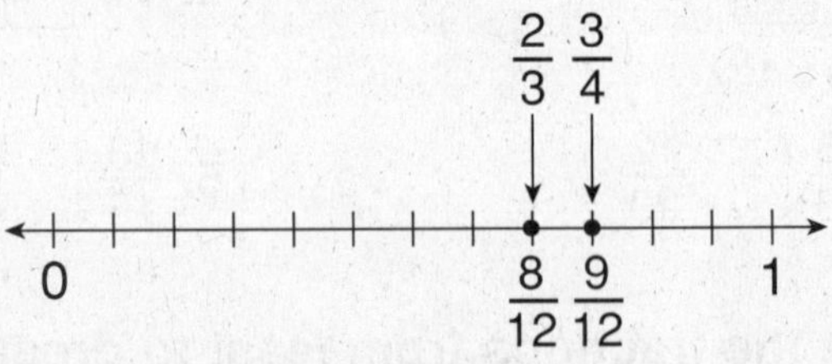

Use the number line to compare the fractions. Write < or >.

1. 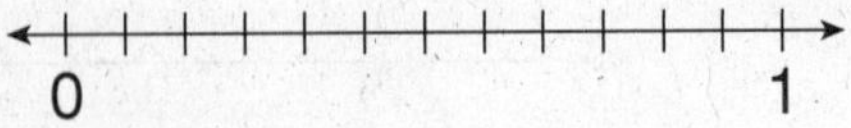

$\frac{1}{3}$ ☐ $\frac{5}{6}$

2. 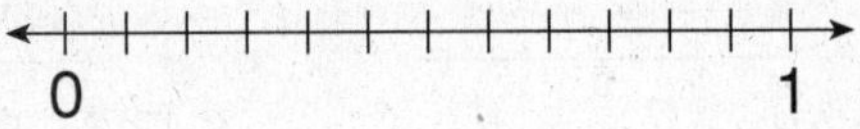

$\frac{5}{10}$ ☐ $\frac{1}{4}$

3. 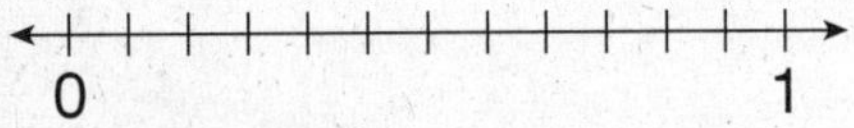

$\frac{11}{12}$ ☐ $\frac{4}{6}$

4. 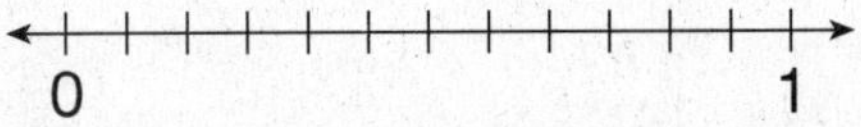

$\frac{6}{9}$ ☐ $\frac{1}{6}$

You can use number lines to order fractions from least to greatest.

To order $\frac{1}{2}$, $\frac{5}{6}$, and $\frac{3}{4}$, graph the values on a number line.

Then list the fractions as they appear from left to right. From least to greatest the fractions are $\frac{1}{2}$, $\frac{3}{4}$, and $\frac{5}{6}$.

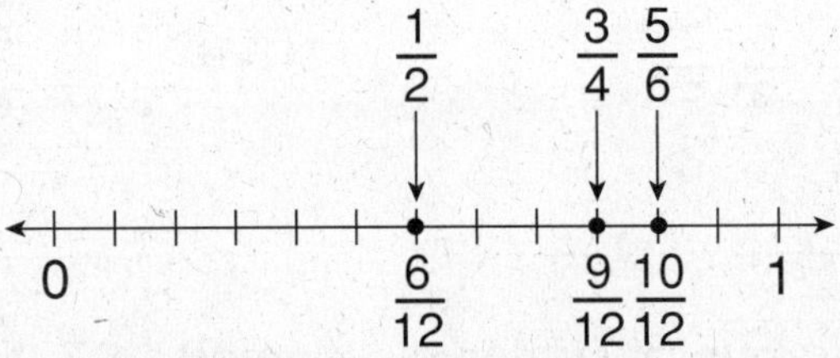

Use the number line to order the fractions from least to greatest.

5. 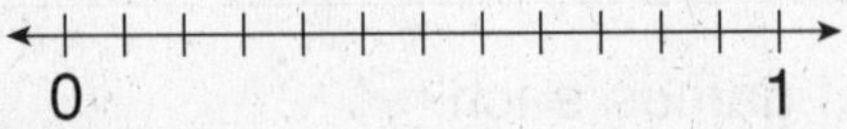

$\frac{7}{12}$, $\frac{1}{4}$, $\frac{4}{8}$ ______________

6. 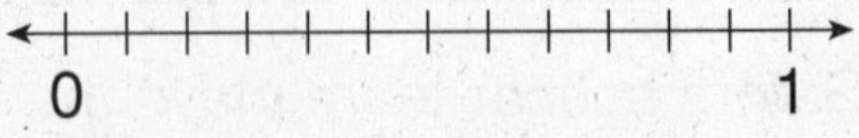

$\frac{2}{3}$, $\frac{7}{12}$, $\frac{2}{8}$ ______________

7. 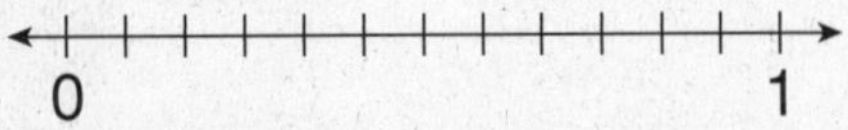

$\frac{1}{2}$, $\frac{3}{4}$, $\frac{3}{9}$ ______________

8. 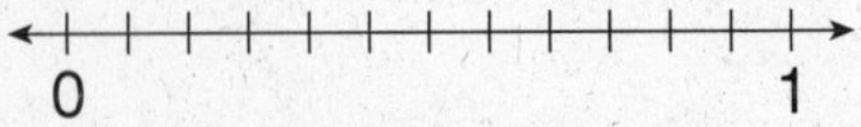

$\frac{5}{12}$, $\frac{1}{3}$, $\frac{5}{6}$ ______________

Name ______________________ Date __________ Class __________

LESSON **4-7**

Challenge

Light as a Feather

The pygmy shrew is the lightest mammal on Earth. It weighs only 0.05 ounces. That's lighter than most feathers! Use the clues below to complete the table with each tiny animal's weight.

Lightest Mammals

Mammal	**Weight** (oz)
	$\frac{1}{20}$
	$\frac{11}{100}$
	$\frac{7}{100}$
	$\frac{9}{50}$
	$\frac{2}{25}$

The smallest pygmy shrews are only $1\frac{1}{2}$ inches long!

1. Thousands of masked shrews live all over North America, but they are rarely seen. Although tiny, they have enormous appetites—often eating more than their body weight each day in bugs, slugs, and other crawly things. A masked shrew weighs more than a kitti's hog-nosed bat, but less than a pipistrelle bat.

2. Several different speices of harvest mice live in the United States. They are excellent climbers and spend much of the night scampering around in search of seeds and bugs to eat. During the day, they stay at home to avoid their worst enemies: hawks, owls, and snakes. A harvest mouse weighs more than any other animal listed in the table.

3. Kitti's hog-nosed bats live only in Asia. They spend most of the day hanging in deep, dark caves. At night, they leave the cave and fly around catching bugs to eat. About the size of a bumblebee, a kitti's hog-nosed bat weighs more than a pygmy shrew, but less than a masked shrew.

4. Pipistrelle bats are the smallest bats in the United States. They are unusual for bats, because they often fly in daylight to catch beetles and other bugs to eat. In winter, they hibernate in caves or other dark places. During that time, the bats are often covered with droplets of water, which sparkle and give them a pearly glow. A pipistrelle bat weighs less than a harvest mouse, but more than a masked shrew.

Name ______________________________ Date ____________ Class ____________

LESSON 4-7

Problem Solving

Comparing and Ordering Fractions

The table shows what fraction of Earth's total land area each of the continents makes up. Use the table to answer the questions.

Earth's Land

Continent	Fraction of Earth's Land
Africa	$\frac{1}{5}$
Antarctica	$\frac{1}{10}$
Asia	$\frac{3}{10}$
Australia	$\frac{1}{20}$
Europe	$\frac{7}{100}$
North America	$\frac{4}{25}$
South America	$\frac{6}{50}$

1. Which continent makes up most of Earth's land?

____

2. Which continent makes up the least part of Earth's land?

____

3. Explain how you would compare the part of Earth's total land area that Australia and Europe make up.

____

____

Circle the letter of the correct answer.

4. Which of these continents covers the greatest part of Earth's total land area?
   A North America
   B South America
   C Europe
   D Australia

5. Which of these continents covers the least part of Earth's total land area?
   F Africa
   G Antarctica
   H Asia
   J Australia

6. Which of the following lists shows the continents written in order from the greatest part of Earth's total land they cover to the least part?
   A Asia, Africa, North America
   B Africa, Asia, North America
   C Asia, South America, North America
   D North America, Asia, South America

7. Which of the following lists shows the continents written in order from the least part of Earth's total land they cover to the greatest part?
   F Antarctica, Europe, South America
   G South America, Antarctica, Europe
   H Australia, Europe, Antarctica
   J Antarctica, Europe, Australia

Name ______________________ Date ____________ Class ____________

LESSON **4-7**

Reading Strategies

Use a Graphic Aid

A number line can help you compare fractions.

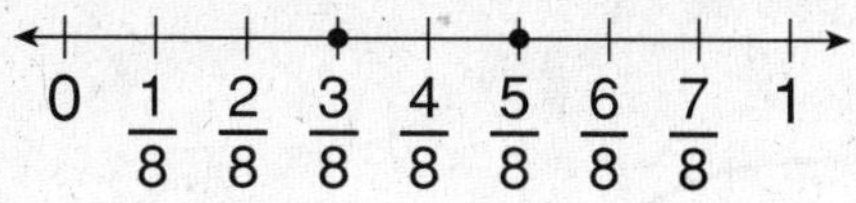

Use the number line and write $>$, $<$, or $=$ for each.

1. $\frac{1}{8} \square \frac{5}{8}$
2. $\frac{5}{8} \square \frac{2}{8}$
3. $\frac{6}{8} \square \frac{4}{8}$

4. When you compare two fractions with the same denominator, how you can tell which one has the greater value?

____

____

A number line can help you compare fractions with different denominators.

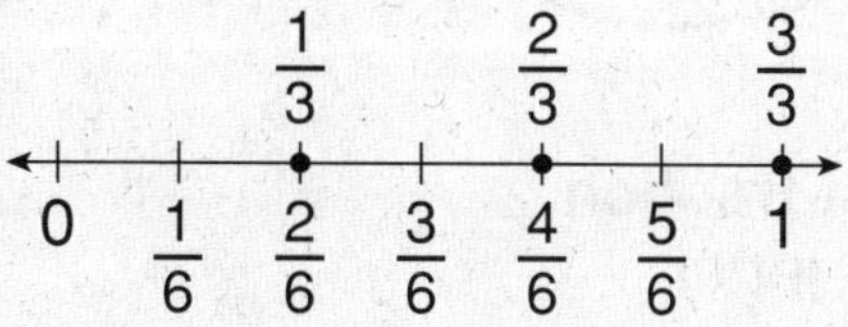

Use the number line and write $>$, $<$, or $=$.

5. $\frac{2}{6} \square \frac{2}{3}$
6. $\frac{5}{6} \square \frac{1}{3}$
7. $\frac{2}{3} \square \frac{4}{6}$
8. $\frac{3}{6} \square \frac{3}{3}$
9. $\frac{5}{6} \square \frac{2}{3}$
10. $\frac{1}{3} \square \frac{1}{6}$

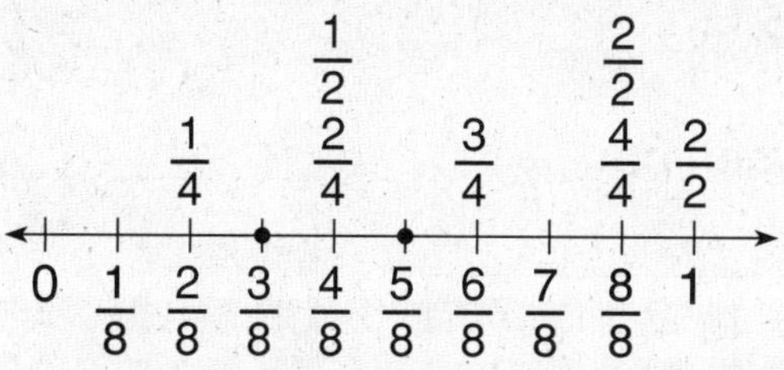

Use the number line and write $>$, $<$, or $=$.

11. $\frac{3}{8} \square \frac{3}{4}$
12. $\frac{3}{4} \square \frac{7}{8}$
13. $\frac{1}{2} \square \frac{5}{8}$
14. $\frac{2}{8} \square \frac{1}{4}$
15. $\frac{2}{4} \square \frac{4}{8}$
16. $\frac{6}{8} \square \frac{1}{2}$

Name ______________________ Date ____________ Class ____________

LESSON 4-7

Puzzles, Twisters & Teasers

Celestial Fractions

Your goal is to pull fractions out of the "sky." Find the single largest fraction and the single smallest fraction in the cloud of fractions below.

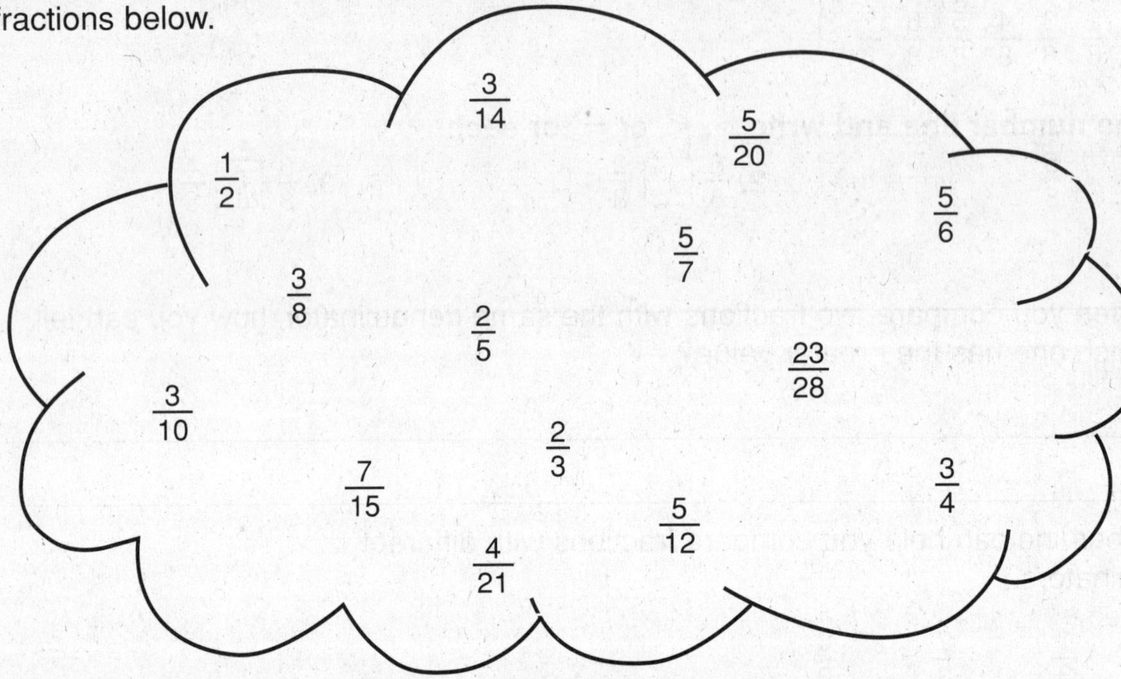

To solve the riddle, take the denominator of the largest fraction as given and find it in the rectangle below. Record the letter directly below that number in the first answer blank. Move clockwise, skip the next letter, and write the letter that comes after that in the next answer blank. Continue skipping and listing letters until you have the eight letters of the first word.

Find the denominator of the smallest fraction and repeat the process to find the second word.

How can you tell that your friend plays cards?

He ___ ___ ___ ___ ___ ___ ___ ___ when he ___ ___ ___ ___ ___!

20 **E**	**14** **S**	**6** **S**	**15** **K**	**8** **H**	**7** **O**
2 **K**					**28** **U**
12 **L**	**5** **L**	**7** **F**	**3** **A**	**10** **F**	**21** **W**

Name ______________________________ Date __________ Class __________

LESSON 4-8

Practice A

Adding and Subtracting with Like Denominators

Subtract. Write each answer in simplest form.

1. $\frac{4}{5} - \frac{1}{5}$ ________

2. $\frac{7}{10} - \frac{5}{10}$ ________

3. $\frac{4}{7} - \frac{3}{7}$ ________

4. $\frac{4}{9} - \frac{1}{9}$ ________

5. $1\frac{5}{6} - 1\frac{2}{6}$ ________

6. $4\frac{7}{8} - 1\frac{4}{8}$ ________

7. $2 - \frac{1}{2}$ ________

8. $1 - \frac{2}{3}$ ________

9. $4 - \frac{1}{4}$ ________

Evaluate each expression for $x = \frac{2}{8}$. Write each answer in simplest form.

10. $x + \frac{3}{8}$ ________

11. $x - \frac{1}{8}$ ________

12. $x + \frac{5}{8}$ ________

13. $\frac{7}{8} - x$ ________

Write each sum or difference in simplest form.

14. $\frac{1}{2} + \frac{1}{2}$ ________

15. $\frac{1}{9} + \frac{3}{9}$ ________

16. $\frac{7}{9} - \frac{1}{9}$ ________

17. $\frac{2}{8} + \frac{4}{8}$ ________

18. $\frac{8}{11} - \frac{4}{11}$ ________

19. $\frac{4}{12} + \frac{4}{12}$ ________

20. Betty read $\frac{2}{3}$ of her book. Use pictures to model how much of her book is left for her to read, and then write your answer in simplest form.

____

21. Sheera ate $\frac{1}{5}$ of the pizza. Andre ate $\frac{2}{5}$ of the same pizza. Use pictures to model how much of the pizza was left over, and then write your answer in simplest form.

____

Name ______________________ Date __________ Class __________

LESSON 4-8

Practice B

Adding and Subtracting with Like Denominators

Subtract. Write each answer in simplest form.

1. $1 - \frac{4}{7}$ ________

2. $\frac{18}{24} - \frac{10}{24}$ ________

3. $2\frac{2}{3} - 1\frac{1}{3}$ ________

4. $8\frac{11}{13} - 5\frac{2}{13}$ ________

5. $5 - 3\frac{1}{4}$ ________

6. $2 - 1\frac{2}{7}$ ________

7. $6\frac{8}{9} - 4\frac{6}{9}$ ________

8. $7\frac{4}{11} - 6\frac{3}{11}$ ________

9. $10 - 5\frac{3}{5}$ ________

Evaluate each expression for $x = \frac{2}{15}$. Write each answer in simplest form.

10. $x + \frac{14}{15}$ ________

11. $x - \frac{1}{15}$ ________

12. $\frac{13}{15} - x$ ________

13. $x + \frac{7}{15}$ ________

Write each sum or difference in simplest form.

14. $\frac{17}{21} - \frac{2}{21}$ ________

15. $\frac{13}{32} + \frac{9}{32}$ ________

16. $\frac{2}{15} + \frac{8}{15}$ ________

17. $27\frac{76}{100} - 14\frac{26}{100}$ ________

18. $\frac{1}{15} + \frac{4}{15} + \frac{5}{15}$ ________

19. $\frac{9}{26} + \frac{2}{26} + \frac{5}{26}$ ________

20. Maria has 8 gallons of paint she wants to use in three rooms of her house. She will use $2\frac{1}{4}$ gallons in the bedroom and $1\frac{1}{4}$ gallons in the bathroom. Use pictures to model how many gallons she will have left to paint the playroom, and then write your answer in simplest form.

____

Name ______________________ Date __________ Class __________

LESSON 4-8

Practice C

Adding and Subtracting with Like Denominators

Subtract. Write each answer in simplest form.

1. $3 - \frac{1}{6}$ ________

2. $7 - \frac{5}{7}$ ________

3. $5 - \frac{3}{8}$ ________

4. $4 - \frac{5}{11}$ ________

5. $5\frac{8}{15} - 3\frac{3}{15}$ ________

6. $8\frac{7}{8} - 4\frac{3}{8}$ ________

7. $3\frac{4}{9} - 2\frac{1}{9}$ ________

8. $9\frac{13}{18} - 4\frac{1}{18}$ ________

9. $100 - 17\frac{9}{16}$ ________

Evaluate each expression for $x = \frac{17}{36}$. Write each answer in simplest form.

10. $\frac{31}{36} - x$ ________

11. $\frac{1}{36} + x$ ________

12. $2\frac{5}{36} + x$ ________

13. $5\frac{25}{36} - x$ ________

Write each sum or difference in simplest form.

14. $\frac{17}{36} + \frac{10}{36}$ ________

15. $\frac{18}{42} + \frac{10}{42}$ ________

16. $\frac{88}{102} - \frac{42}{102}$ ________

17. $48\frac{11}{18} - 37\frac{2}{18}$ ________

18. $\frac{19}{20} + \frac{16}{20} + \frac{13}{20}$ ________

19. $\frac{9}{27} + \frac{3}{27} + \frac{3}{27}$ ________

20. Shelby practiced piano for $1\frac{1}{8}$ hours on Monday. On Tuesday she practiced for $\frac{5}{8}$ hour, and on Wednesday for $1\frac{3}{8}$ hours. Use pictures to model how many hours she should practice on Thursday if she wants to practice 5 hours this week, and then write your answer in simplest form.

____

21. What are the next two terms in the following pattern: $2\frac{1}{4}$, $3\frac{1}{2}$, $4\frac{3}{4}$? Write them in simplest form.

____

Name ______________________ Date ____________ Class ____________

LESSON 4-8

Reteach

Adding and Subtracting with Like Denominators

You can use fraction strips to add like fractions.

To find $\frac{1}{8} + \frac{3}{8}$, first model the expression using $\frac{1}{8}$ fraction pieces.

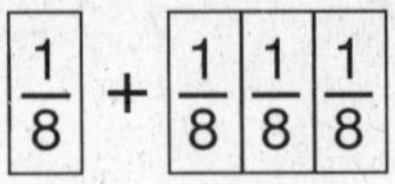

The total number of fraction pieces is 4.

So, $\frac{1}{8} + \frac{3}{8} = \frac{4}{8}$.

Then write the sum in simplest form. $\frac{4}{8} = \frac{1}{2}$.

Write each sum in simplest form.

1. $\frac{1}{4} + \frac{1}{4}$ ____________
2. $\frac{1}{5} + \frac{3}{5}$ ____________
3. $\frac{1}{6} + \frac{5}{6}$ ____________
4. $\frac{5}{12} + \frac{1}{12}$ ____________

You can use fraction strips to subtract like fractions.

To find $\frac{5}{6} - \frac{1}{6}$, use fraction strips to model the first fraction.

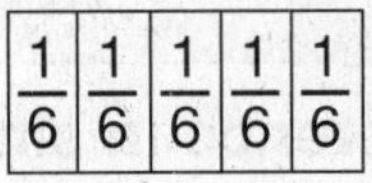

Next, take away fraction strips that represent the second fraction.

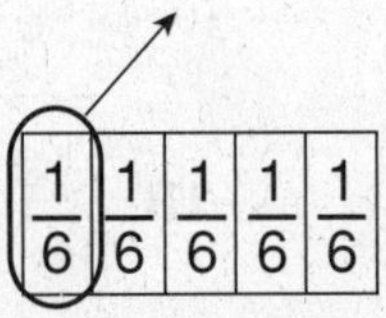

The remaining fraction strips represent the difference.

$\frac{5}{6} - \frac{1}{6} = \frac{4}{6}$

$\frac{4}{6} = \frac{2}{3}$ in simplest form.

Write each difference in simplest form.

5. $\frac{5}{8} - \frac{3}{8}$ ____________
6. $\frac{7}{12} - \frac{5}{12}$ ____________
7. $\frac{3}{4} - \frac{1}{4}$ ____________
8. $\frac{9}{10} - \frac{3}{10}$ ____________

Name ______________________ Date ____________ Class ____________

LESSON **4-8**

Challenge

Adding Equals Subtracting?

Use one expression from Box A and one from Box B to write an equation in the pans below. The two expressions must be equal to keep the scale balanced. Use each expression only once.

A		B	
$\frac{1}{4} + \frac{2}{4}$	$\frac{1}{27} + \frac{2}{27}$	$\frac{11}{7} - \frac{6}{7}$	$\frac{11}{12} - \frac{1}{12}$
$\frac{1}{6} + \frac{4}{6}$	$\frac{2}{7} + \frac{3}{7}$	$\frac{4}{8} - \frac{1}{8}$	$1 - \frac{8}{9}$
$\frac{3}{16} + \frac{3}{16}$	$\frac{1}{5} + \frac{1}{5}$	$1 - \frac{4}{10}$	$2 - 1\frac{3}{5}$
$\frac{1}{18} + \frac{2}{18}$	$\frac{4}{15} + \frac{5}{15}$	$2 - 1\frac{1}{4}$	$\frac{5}{6} - \frac{4}{6}$

1.

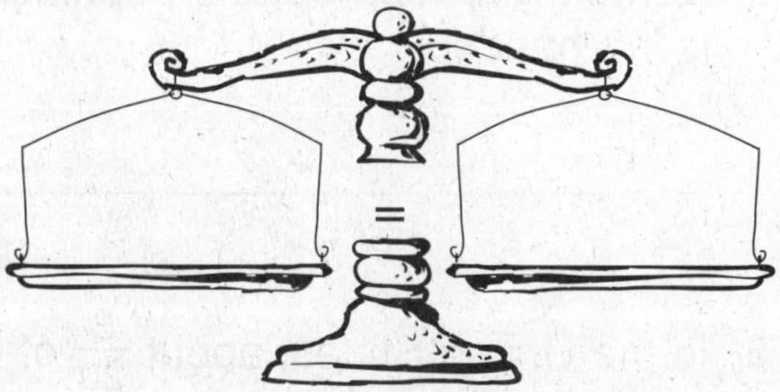

2.

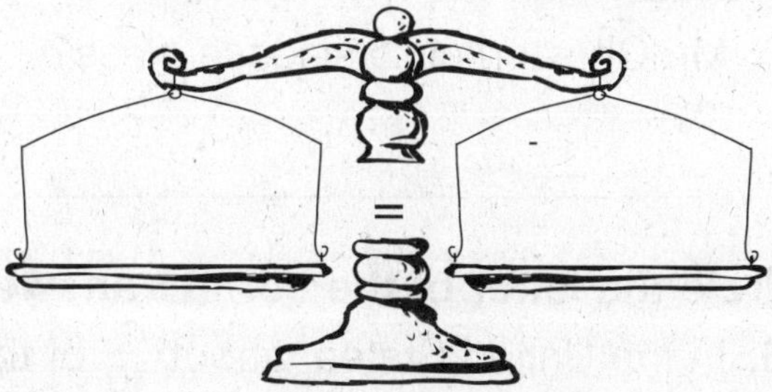

3.

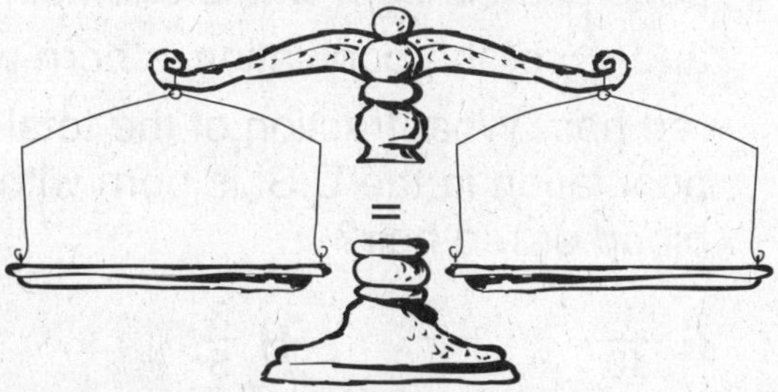

4.

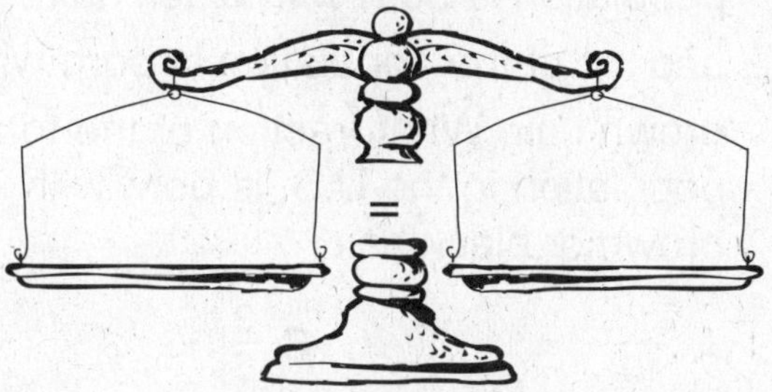

5.

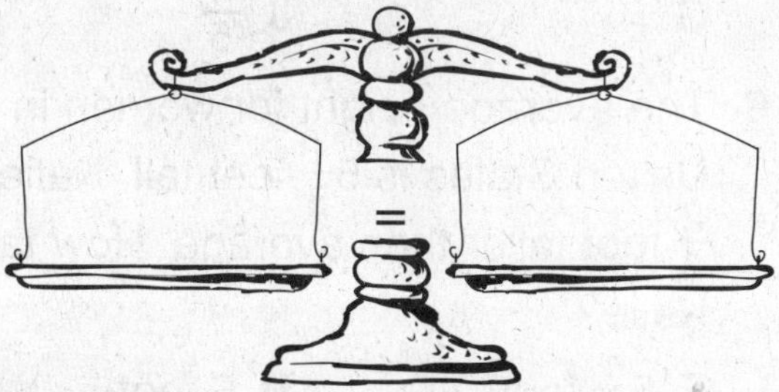

6.

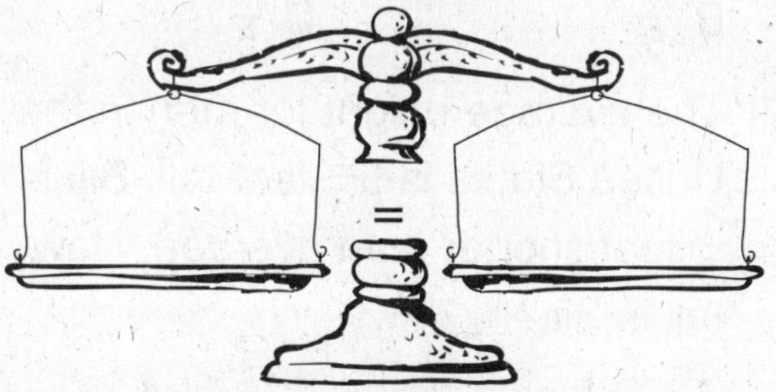

7.

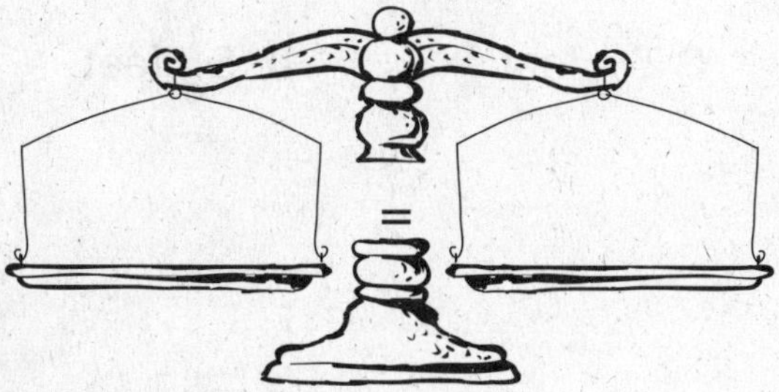

8.

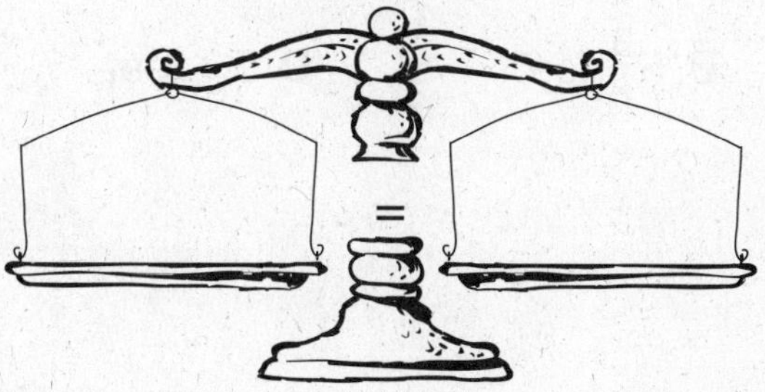

Name ______________________ Date ____________ Class ____________

LESSON 4-8

Problem Solving

Adding and Subtracting with Like Denominators

Write the answers in simplest form.

1. About $\frac{3}{10}$ of Earth's surface is covered by land, and the rest is water. What fraction of Earth's surface is covered by water?

____

2. A recipe for cookies calls for $\frac{3}{8}$ cup of chocolate chips. Tameeka wants to double the recipe. How much chocolate chips will she use?

____

3. In Mr. Chesterfield's science class, $\frac{2}{9}$ of the boys like his class. Three times as many girls like his science class. How many girls like Mr. Chesterfield's science class?

____

4. In the United States, $\frac{6}{50}$ of the population is left-handed men and $\frac{5}{50}$ of the population is left-handed women. What part of the population is left-handed?

____

Circle the letter of the correct answer.

5. In the United States, about $\frac{1}{10}$ of the population is born with black hair, and $\frac{7}{10}$ of the population is born with brown hair. What fraction of the total population in the U.S. is born with brown or black hair?

A $\frac{1}{10}$ **C** $\frac{3}{5}$

B $\frac{1}{5}$ **D** $\frac{4}{5}$

6. In the United States, about $\frac{3}{20}$ of the population is born with blond hair, and $\frac{1}{20}$ of the population is born with red hair. What fraction of the total population in the U.S. is born with blond or red hair?

F $\frac{1}{10}$ **H** $\frac{3}{5}$

G $\frac{1}{5}$ **J** $\frac{4}{5}$

7. The average height for men in the United States is $5\frac{2}{3}$ feet tall. Bill is $\frac{1}{3}$ foot shorter than average. How tall is Bill?

A $5\frac{1}{3}$ feet **C** 6 feet

B $5\frac{3}{6}$ feet **D** $5\frac{1}{6}$ feet

8. The average height for women in the United States is $5\frac{1}{3}$ feet tall. Katie is $\frac{2}{3}$ foot taller than average. How tall is Katie?

F $5\frac{1}{3}$ feet **H** 6 feet

G $5\frac{3}{6}$ feet **J** $5\frac{1}{6}$ feet

Name ______________________________ Date ____________ Class ____________

LESSON 4-8

Reading Strategies

Use a Visual Model

To add like fractions, add the numerators and simplify.

These rectangles show an addition problem with like fractions.

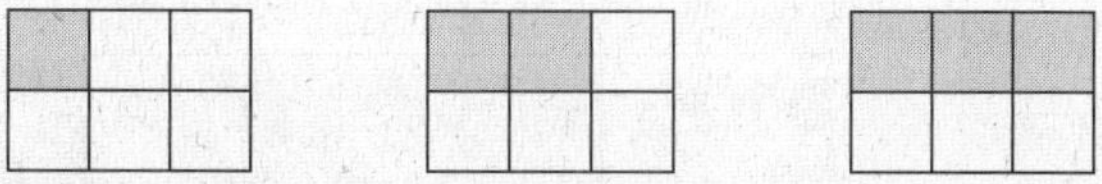

one sixth + two sixths = three sixths

$$\frac{1}{6} + \frac{2}{6} = \frac{3}{6}$$

$$= \frac{1}{2}$$

Use the model to answer the questions.

1. What is one-sixth plus one-sixth? ______________________

2. What is $\frac{2}{6} + \frac{2}{6}$? ______________________

3. What is three-sixths plus two-sixths? ______________________

To subtract like fractions, subtract the numerators and simplify.

These rectangles show a subtraction problem with like fractions.

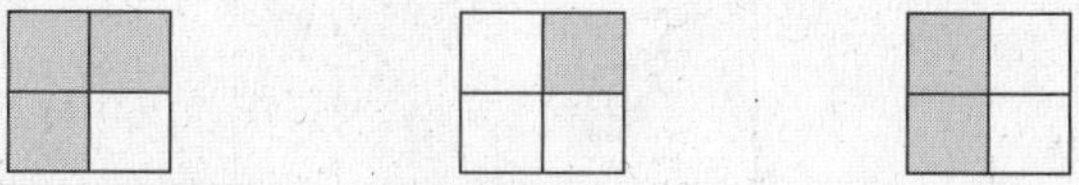

three fourths − one fourth = two fourths

$$\frac{3}{4} - \frac{1}{4} = \frac{2}{4}$$

$$= \frac{1}{2}$$

Use the rectangles to answer each question.

4. What is $\frac{2}{4} - \frac{1}{4}$? ______________________

5. What is four-fourths minus one-fourth? ______________________

6. What is $\frac{3}{4} - \frac{2}{4}$? ______________________

Name ______________________ Date ____________ Class ____________

LESSON 4-8

Puzzles, Twisters & Teasers

Magic Squares

In a magic square every row, column, and diagonal has the same sum. Identify which of the squares below are magic. Then take the letters in the four corners of the squares that are magic and unscramble them to solve the riddle.

T 6	$2\frac{1}{2}$	D 5
$3\frac{1}{2}$	$4\frac{1}{2}$	$5\frac{1}{2}$
L 4	$6\frac{1}{2}$	Y 3

Sum: __________

Magic? __________

O $10\frac{1}{8}$	$8\frac{3}{8}$	A $9\frac{5}{8}$
$8\frac{7}{8}$	$9\frac{3}{8}$	$9\frac{7}{8}$
A $9\frac{1}{8}$	$10\frac{3}{8}$	L $8\frac{5}{8}$

Sum: __________

Magic? __________

B 7	$2\frac{1}{4}$	G 5
$3\frac{1}{4}$	$4\frac{1}{4}$	$5\frac{2}{4}$
N 4	$6\frac{3}{4}$	I 3

Sum: __________

Magic? __________

X 6	$5\frac{2}{3}$	S 5
$7\frac{1}{3}$	$4\frac{2}{3}$	8
K 2	$6\frac{1}{3}$	E 3

Sum: __________

Magic? __________

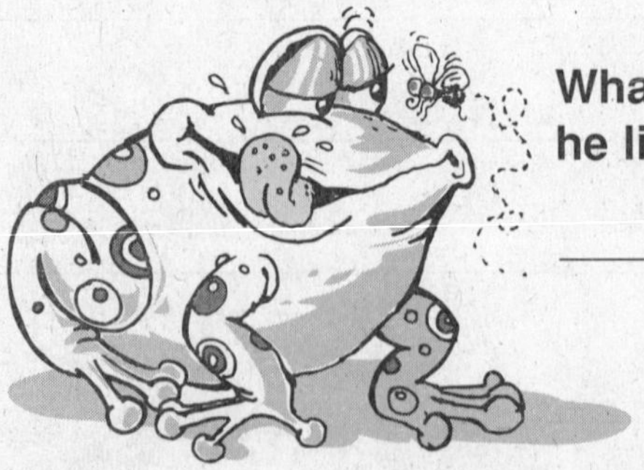

What did the student say when the teacher asked him if he liked studying frogs in science?

___ ___ ___ ___ ‑___ ___ ___ ___ ___

Name ________________________ Date ____________ Class ____________

LESSON 4-9

Practice A

Estimating Fraction Sums and Differences

Round each number to 0, $\frac{1}{2}$, or 1.

1. $\frac{1}{6}$ ____________ **2.** $\frac{3}{7}$ ____________ **3.** $\frac{7}{8}$ ____________

4. $\frac{2}{5}$ ____________ **5.** $\frac{9}{10}$ ____________ **6.** $\frac{2}{15}$ ____________

Estimate each sum or difference by rounding to 0, $\frac{1}{2}$, or 1.

7. $\frac{2}{3} + \frac{3}{4}$ ____________ **8.** $\frac{5}{6} - \frac{3}{5}$ ____________ **9.** $\frac{4}{9} + \frac{1}{8}$ ____________

10. $\frac{8}{9} - \frac{6}{7}$ ____________ **11.** $\frac{1}{4} + \frac{2}{3}$ ____________ **12.** $\frac{3}{4} - \frac{2}{3}$ ____________

13. $\frac{4}{7} + \frac{3}{5}$ ____________ **14.** $\frac{1}{5} + \frac{4}{9}$ ____________ **15.** $\frac{3}{4} - \frac{4}{7}$ ____________

Use the table for Exercises 16 and 17.

16. About how far did Mia swim during week 1 and week 2 altogether?

____

17. About how much farther did Mia swim during week 3 than during week 1?

____

Mia's Swimming Distances

Week	**Distance** (mi)
1	$1\frac{1}{4}$
2	$\frac{2}{3}$
3	$1\frac{5}{6}$

18. Shelley used $\frac{3}{5}$ ounce of water in her experiment. Ali used $\frac{7}{9}$ ounce of water in his experiment. Who used more water and about how much more water was used?

____

19. Fred ran $1\frac{6}{11}$ of a mile, and then he walked $1\frac{5}{8}$ of a mile. About how many miles did Fred cover in all?

____

Name ______________________ Date ____________ Class ____________

LESSON **4-9**

Practice B

Estimating Fraction Sums and Differences

Estimate each sum or difference by rounding to 0, $\frac{1}{2}$, or 1.

1. $\frac{5}{6} + \frac{3}{10}$ ______

2. $\frac{7}{9} - \frac{4}{5}$ ______

3. $\frac{9}{10} - \frac{3}{7}$ ______

4. $\frac{4}{9} + \frac{1}{4}$ ______

5. $\frac{1}{8} + \frac{1}{6}$ ______

6. $\frac{7}{8} - \frac{4}{5}$ ______

7. $\frac{5}{8} + \frac{2}{7}$ ______

8. $\frac{7}{10} + \frac{11}{12}$ ______

9. $\frac{8}{9} + \frac{4}{7}$ ______

10. $\frac{5}{11} + 1\frac{2}{7}$ ______

11. $1\frac{6}{11} - \frac{2}{5}$ ______

12. $4\frac{2}{7} - 1\frac{7}{9}$ ______

Use the table for Exercises 13–15.

Fruit Punch

Ingredient	**Amount** (cups)
Orange juice	$4\frac{3}{5}$
Cranberry juice	$2\frac{1}{4}$
Ginger ale	$\frac{7}{8}$

13. About how much more orange juice than ginger ale is used in the punch?

______

14. About how much juice is used in the punch?

______

15. About how many cups of fruit punch does this recipe make?

______

16. Damonte rolled the medicine ball $9\frac{3}{4}$ feet. Zachary rolled it $9\frac{7}{12}$ feet. Who rolled the medicine ball the farthest? About how much farther?

______

17. Sara ran $5\frac{6}{7}$ miles on Monday and $4\frac{1}{4}$ miles on Tuesday. About how many miles did she run in all during those two days?

______

Name ______________________ Date ____________ Class ____________

LESSON 4-9

Practice C

Estimating Fraction Sums and Differences

Estimate each sum or difference by rounding to 0, $\frac{1}{2}$, or 1.

1. $\frac{8}{12} + \frac{5}{11}$ ____________

2. $\frac{4}{10} - \frac{2}{7}$ ____________

3. $\frac{3}{9} + \frac{3}{5}$ ____________

4. $\frac{7}{12} - \frac{8}{15}$ ____________

5. $\frac{4}{9} + \frac{7}{13}$ ____________

6. $\frac{19}{20} - \frac{7}{9}$ ____________

7. $\frac{10}{12} - \frac{5}{8} + \frac{3}{4}$ ____________

8. $\frac{8}{11} + \frac{9}{13} + \frac{5}{11}$ ____________

9. $\frac{6}{9} + \frac{7}{12} - \frac{2}{3}$ ____________

10. $1\frac{2}{9} + 2\frac{3}{17}$ ____________

11. $7\frac{5}{6} - \frac{8}{9}$ ____________

12. $4\frac{8}{15} - 3\frac{2}{7}$ ____________

Estimate each sum or difference to compare. Write $<$ or $>$.

13. $1\frac{5}{11} + 2\frac{6}{7}$ ☐ 4

14. $3\frac{1}{4} - \frac{3}{5}$ ☐ 2

15. $2\frac{4}{9} - 1\frac{5}{11}$ ☐ 2

16. $5\frac{3}{7} + 3\frac{1}{4}$ ☐ 9

17. $4\frac{6}{13} - 3\frac{2}{3}$ ☐ 1

18. $3\frac{3}{10} + 2\frac{5}{14}$ ☐ 4

19. Whitney's candy bag weighed $2\frac{7}{20}$ pounds. Bobby's weighed $2\frac{5}{9}$ pounds. About how much more did Bobby's candy weigh? About how much candy did they buy all together?

____

20. Jamie walked $3\frac{5}{7}$ miles on Monday, $4\frac{3}{11}$ miles on Tuesday, and $2\frac{5}{11}$ miles on Wednesday. About how many miles did she walk all together? Her goal was to walk 12 miles. About how much over or under her goal was she?

____

Name ______________________ Date ____________ Class ____________

LESSON **4-9**

Reteach

Estimating Fraction Sums and Differences

You can use number lines to help you estimate fraction sums and differences.

To estimate the sum of $\frac{5}{6}$ and $\frac{1}{3}$, locate each fraction on a number line. Then round each fraction to 0, $\frac{1}{2}$, or 1.

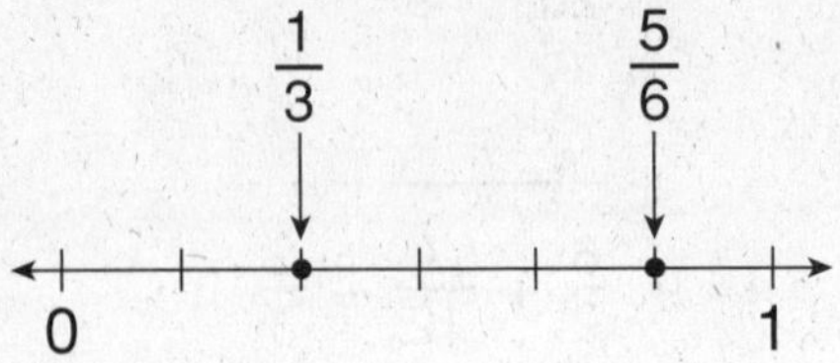

$\frac{5}{6} + \frac{1}{3} \approx$

$1 + \frac{1}{2} = 1\frac{1}{2}$

So, $\frac{5}{6} + \frac{1}{3}$ is about $1\frac{1}{2}$.

To estimate the difference between $\frac{7}{8}$ and $\frac{2}{4}$, locate each fraction on a number line. Then round each fraction to 0, $\frac{1}{2}$, or 1.

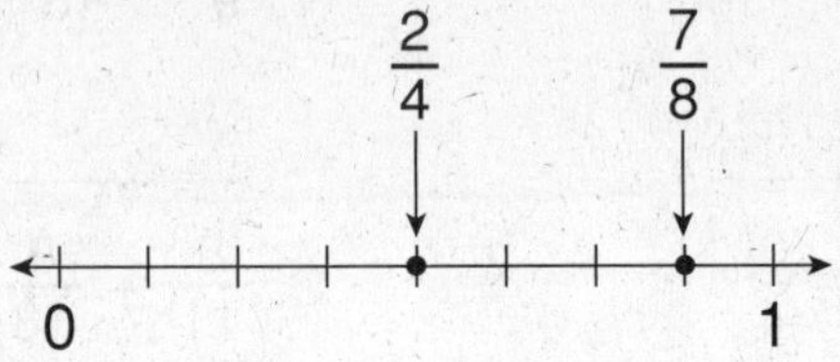

$\frac{7}{8} - \frac{2}{4} \approx$

$1 - \frac{1}{2} = \frac{1}{2}$

So, $\frac{7}{8} - \frac{2}{4}$ is about $\frac{1}{2}$.

Use the number line to round each fraction to 0, $\frac{1}{2}$, or 1 to estimate each sum or difference.

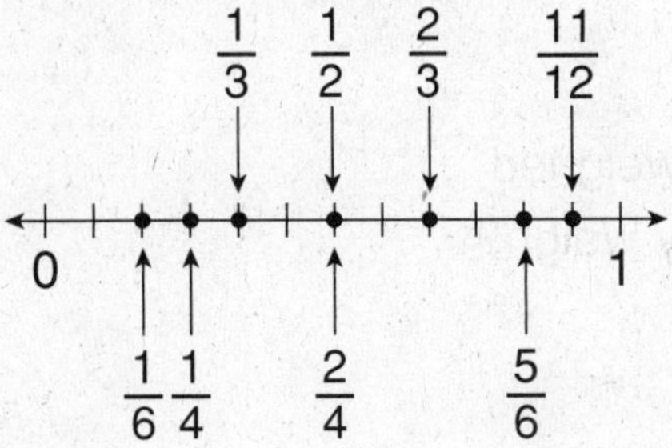

1. $\frac{5}{6} + \frac{1}{6}$ ________

2. $\frac{11}{12} - \frac{1}{2}$ ________

3. $\frac{2}{3} + \frac{2}{4}$ ________

4. $\frac{1}{4} - \frac{1}{3}$ ________

Use the number line to round each fraction to 0, $\frac{1}{2}$, or 1 to estimate each sum or difference.

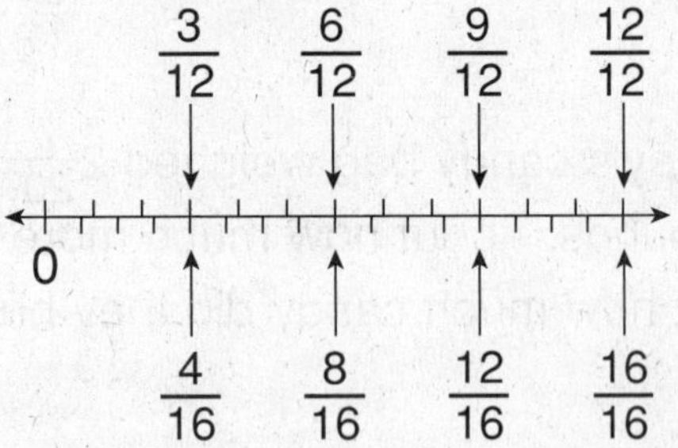

5. $\frac{7}{12} + \frac{2}{6}$ ________

6. $\frac{5}{6} - \frac{3}{8}$ ________

7. $\frac{1}{4} + \frac{2}{6}$ ________

8. $\frac{7}{8} + \frac{14}{16}$ ________

Name ______________________ Date __________ Class __________

LESSON **4-9**

Challenge

What's on the Menu?

The menu below shows the average amounts of food most Americans eat each day. Use the menu to complete the diet estimation chart that follows.

Average American Daily Menu

Dairy Products: $1\frac{2}{5}$ **pounds**

Processed Fruits and Vegetables: $1\frac{1}{20}$ **pounds**

Flour and Cereal Products: $\frac{27}{50}$ **pound**

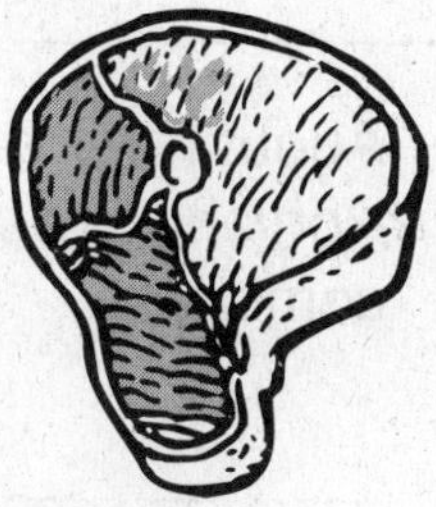

Red Meat: $\frac{9}{25}$ **pound**

Sweetener Products: $\frac{21}{50}$ **pound**

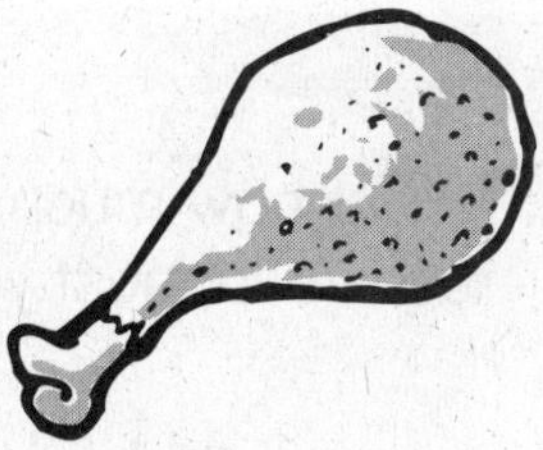

Poultry: $\frac{9}{50}$ **pounds**

Fresh Fruits and Vegetables: $\frac{9}{10}$ **pounds**

Fats and Oils: $\frac{9}{50}$ **pounds**

Estimated American Diet

Foods	Estimated Daily Consumption
Red meat and poultry	
Fruits and vegetables	
Sweetener products and fats and oils	
Dairy and flour and cereal products	
Total Daily Food Consumption:	

Name ______________________ Date __________ Class __________

LESSON 4-9

Problem Solving

Estimating Fraction Sums and Differences

Use the table to answer the questions.

Portland, Oregon, Average Monthly Rainfall

Month	Jan	Feb	Mar	Apr	May	Jun	Jul	Aug	Sep	Oct	Nov	Dec
Rain (in.)	$5\frac{2}{5}$	$3\frac{9}{10}$	$3\frac{3}{5}$	$2\frac{2}{5}$	$2\frac{1}{10}$	$1\frac{1}{2}$	$\frac{3}{5}$	$1\frac{1}{10}$	$1\frac{4}{5}$	$2\frac{7}{10}$	$5\frac{3}{10}$	$6\frac{1}{10}$

1. About how much does it rain in Portland in January and February combined?

____

2. About how much more does it rain in Portland in October than in September?

____

3. In most years, about how much rain does Portland receive from May through July?

____

4. What is the difference between Portland's average rainfall in March and May?

____

Circle the letter of the correct answer.

5. What is the difference in rainfall between Portland's rainiest and driest months?

A about $2\frac{1}{2}$ inches

B about 5 inches

C about $6\frac{1}{2}$ inches

D about $7\frac{1}{2}$ inches

6. About how much rain does Portland receive in most years all together?

F about $25\frac{1}{2}$ inches

G about $30\frac{1}{2}$ inches

H about $32\frac{1}{2}$ inches

J about $36\frac{1}{2}$ inches

7. About how much rain does Portland receive during its three rainiest months all together?

A about 17 inches

B about 16 inches

C about 18 inches

D about 15 inches

8. In which month in Portland can you expect about $\frac{1}{2}$ inch less rainfall than in June?

F May

G July

H September

J August

Name ______________________ Date __________ Class __________

LESSON 4-9

Reading Strategies

Use a Graphic Aid

When you don't need exact values of fractions, you can use a number line to help you estimate the values by rounding.

Fractions Close to 1

The number line shows that the fractions $\frac{5}{6}$ and $\frac{7}{8}$ are both close to 1. When the numerator and denominator of a fraction are close to the same value, round to 1.

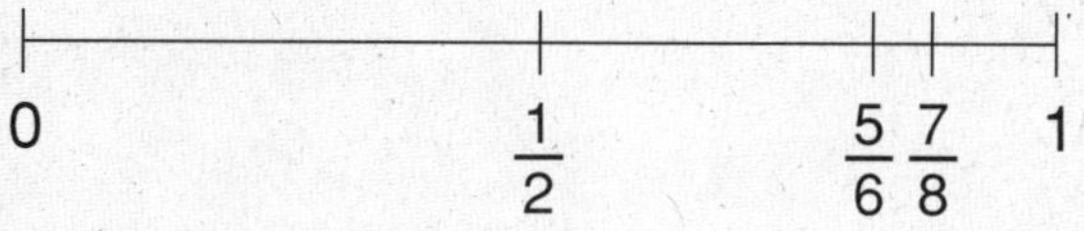

Fractions Close to $\frac{1}{2}$

The number line shows fractions such as $\frac{6}{13}$, $\frac{9}{20}$, and $\frac{12}{23}$ that are close to $\frac{1}{2}$. When the numerator is about half the value of the denominator, round to $\frac{1}{2}$.

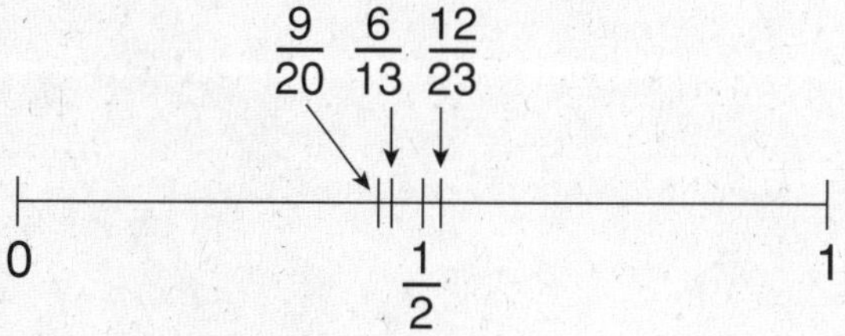

Fractions Close to 0

The number line shows fractions such as $\frac{1}{15}$, $\frac{6}{50}$. and $\frac{2}{25}$ that are close to 0. When the numerator is much less than the denominator, round to 0.

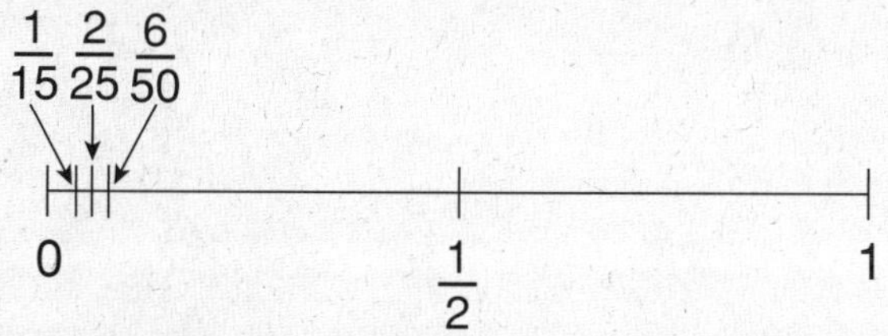

Estimate the value of each fraction. Write *close to 0, close to $\frac{1}{2}$*, or *close to 1.*

1. $\frac{12}{25}$ ______________________ **2.** $\frac{1}{17}$ ______________________

3. $\frac{8}{9}$ ______________________ **4.** $\frac{9}{10}$ ______________________

Name ______________________ Date __________ Class __________

LESSON **4-9**

Puzzles, Twisters & Teasers

Call to Arms

Estimate the sum or difference by rounding each fraction or mixed number. Circle your estimated answers.

To solve the riddle, fill in the blanks with the matching letters.

1. About how much is $\frac{7}{8} + \frac{5}{6}$?

R 2 **U** 3 **W** 4

2. About how much is $6\frac{2}{3} - 2\frac{1}{4}$?

K 3 **D** 4 **S** 5

3. About how much is $2\frac{5}{7} + 3\frac{1}{4}$?

H 5 **E** 6 **C** 7

4. About how much is $8\frac{1}{4} - 7\frac{5}{6}$?

B 2 **T** 1 **A** 0

5. About how much is $4\frac{3}{8} + \frac{4}{7}$?

M 5 **F** 6 **P** 4

How does an octopus go to war? ___ ___ ___ ___ ___
4 1 5 3 2

LESSON 4-1

Practice A

Divisibility

Tell whether each number is even or odd.

1. 10 — even
2. 35 — odd
3. 20 — even
4. 14 — even
5. 17 — odd
6. 19 — odd

Tell whether each number is divisible by 2, 3, 4, 5, 6, 9, and 10.

7. 32 — 2; 4
8. 81 — 3; 9
9. 45 — 3; 5; 9
10. 56 — 2; 4
11. 21 — 3
12. 72 — 2; 3; 4; 6; 9

Tell whether each number is prime or composite.

13. 4 — composite
14. 9 — composite
15. 23 — prime
16. 11 — prime
17. 12 — composite
18. 16 — composite

19. Three brothers bought a pack of 24 baseball cards to share. Can they each get the same number of cards? How do you know?

Yes; because 24 is divisible by 3

20. Five friends decide to order a pizza. If the pizza is cut into 8 slices, will they each be able to get the same number of slices without any pizza left over? How do you know?

No; because 8 is not divisible by 5

LESSON 4-1

Practice B

Divisibility

Tell whether each number is divisible by 2, 3, 4, 5, 6, 9, and 10.

1. 90 — 2; 3; 5; 6; 9; 10
2. 416 — 2; 4
3. 308 — 2; 4
4. 540 — 2; 3; 4; 5; 6; 9; 10
5. 804 — 2; 3; 4; 6
6. 225 — 3; 5; 9
7. 663 — 3
8. 972 — 2; 3; 4; 6; 9
9. 836 — 2; 4

Tell whether each number is prime or composite.

10. 33 — composite
11. 69 — composite
12. 41 — prime
13. 45 — composite
14. 58 — composite
15. 87 — composite
16. 61 — prime
17. 53 — prime
18. 99 — composite

19. Dan counted all the coins in his bank, and he had 72 quarters. Can he exchange the quarters for an even amount of dollar bills? How do you know?

Yes; because there are 4 quarters in 1 dollar, and 72 is divisible by 4.

20. A small town purchased 196 American flags for its Memorial Day parade. Eight locations were selected to display the flags. Can each location have the same number of flags? If no, explain why not. If yes, how many flags will be displayed at each location?

No; because 196 is not divisible by 8.

LESSON 4-1

Practice C

Divisibility

Tell whether each number is divisible by 2, 3, 4, 5, 6, 9, and 10.

1. 1,524 — 2; 3; 4; 6
2. 1,000 — 2; 4; 5; 10
3. 4,455 — 3; 5; 9
4. 2,160 — 2; 3; 4; 5; 6; 9; 10
5. 3,174 — 2; 3; 6
6. 8,433 — 3; 9
7. 5,745 — 3; 5
8. 6,200 — 2; 4; 5; 10
9. 13,000 — 2; 4; 5; 10

Write all the prime numbers between each set of numbers.

10. 1–10 — 2; 3; 5; 7
11. 21–30 — 23; 29
12. 35–45 — 37; 41; 43
13. 50–60 — 53; 59
14. 65–75 — 67; 71; 73
15. 75–100 — 79; 83; 89; 97

Replace each question mark with a digit that will make the number divisible by 9.

16. 35? — 1
17. 2,68? — 2
18. 7?4 — 7
19. 53,?69 — 4

20. Several Boy Scouts worked together to create a line of 4,900 dominoes. They each placed the same amount of dominoes in the line. There were between 50 and 100 Boy Scouts working on the line. What are all the possible numbers of Boy Scouts who could have created the domino line?

50, 70, 98, and 100

21. Use the divisibility rules you know to write a divisibility rule for 12.

Possible answer: If a number is divisible by 3 and 4, then it is divisible by 12.

LESSON 4-1

Reteach

Divisibility

A number is divisible by another number if the quotient is a whole number with no remainder.

15 is divisible by 3 because $15 \div 3 = 5$. There are rules to help you figure out if a number is divisible by another number.

<u>Divisibility Rules</u>

A number is divisible by:

- 2 if the last digit is even.
- 3 if the sum of the digits is divisible by 3.
- 4 if the last two digits form a number that is divisible by 4.
- 5 if the last digit is 0 or 5.
- 6 if the number is divisible by both 2 and 3.
- 9 if the sum of the digits is divisible by 9.
- 10 if the last digit is 0.

To tell whether 315 is divisible by 2, 3, 4, 5, 6, 9, and 10, you can use the divisibility rules listed above.

5 is not an even number, so 315 **is not divisible** by 2.
$3 + 1 + 5 = 9$. 9 is divisible by 3, so 315 **is divisible** by 3.
15 is not divisible by 4, so 315 **is not divisible** by 4.
The last digit is 0 or 5, so 315 **is divisible** by 5.
315 is divisible by 3, but not 2. So, 315 **is not divisible** by 6.
The sum of the digits is 9 and 9 is divisible by 9. So 315 **is divisible** by 9.
The last digit is not 0, so 315 **is not divisible** by 10.

315 is divisible by 3, 5, and 9.

Use divisibility rules to tell whether each number is divisible by 2, 3, 4, 5, 6, 9, and 10.

1. 120 — 2; 3; 4; 5; 6; 10
2. 435 — 3; 5
3. 228 — 2; 3; 4; 6
4. 540 — 2; 3; 4; 5; 6; 9; 10
5. 144 — 2; 3; 4; 6; 9
6. 634 — 2
7. 402 — 2; 3; 6
8. 320 — 2; 4; 5; 10

LESSON **4-1**
Reteach
Divisibility (continued)

Numbers that are divisible by more than 2 numbers are called composite numbers.

12 is a composite number because it is divisible by 1, 2, 3, 4, 6, and 12.

$12 \div 1 = 12$ $12 \div 2 = 6$ $12 \div 3 = 4$

$12 \div 4 = 3$ $12 \div 6 = 2$ $12 \div 12 = 1$

Numbers that are divisible by exactly 2 numbers, 1 and itself, are called prime numbers.

5 is a prime number because it is divisible by only 1 and 5.

$5 \div 1 = 5$ $5 \div 5 = 1$

Tell whether each number is prime or composite.

9. 7 — prime
10. 12 — composite
11. 17 — prime
12. 28 — composite

13. 37 — prime
14. 45 — composite
15. 79 — prime
16. 88 — composite

17. 72 — composite
18. 59 — prime
19. 94 — composite
20. 21 — composite

21. 47 — prime
22. 55 — composite
23. 67 — prime
24. 81 — composite

LESSON **4-1**
Challenge
The Chinese Calendar

The Chinese calendar runs in cycles of 12 years. Each year in a cycle is named after an animal. The Chinese believe the animal ruling the year in which you were born has a great effect on your personality. They say, "This is the animal that hides in your heart."

2000 was a year of the dragon.

The Chinese calendar cycle below is for the years 1983–1994. Use the clues to write the last two digits in the year that goes with each animal. Then read the character traits associated with the animals. Which animal hides in your heart?

Tiger: 19 86 **Divisible by:** 2 and 43 **Traits:** bold, adventurous	**Sheep:** 19 91 **Divisible by:** 7 **Traits:** charming, artistic	**Pig:** 19 83 **Divisible by:** 1 and 83 **Traits:** loyal, tolerant
Dog: 19 94 **Divisible by:** 2 and 47 **Traits:** honest, faithful	**Dragon:** 19 88 **Divisible by:** 2, 4, and 8 **Traits:** fun, energetic	**Rat:** 19 84 **Divisible by:** 4 and 6 **Traits:** generous, creative
Horse: 19 90 **Divisible by:** 5 and 9 **Traits:** friendly, hardworking	**Rooster:** 19 93 **Divisible by:** 3 and 31 **Traits:** decisive, proud	**Ox:** 19 85 **Divisible by:** 5 **Traits:** confident, leaderlike
Snake: 19 89 **Divisible by:** 1 and 89 **Traits:** charming, wise	**Monkey:** 19 92 **Divisible by:** 2 and 4 **Traits:** smart, likable	**Rabbit:** 19 87 **Divisible by:** 3 and 29 **Traits:** friendly, cooperative

LESSON **4-1**
Problem Solving
Divisibility

Use the table to answer the questions.

Subways Around the World

City, Country	Length (mi)
New York, U.S.	247
Mexico City, Mexico	111
Paris, France	125
Moscow, Russia	152
Seoul, South Korea	83
Tokyo, Japan	105

1. Which city's subway has a length that is a prime number of miles?

   Seoul, South Korea

2. Which subway could be evenly broken into sections of 2 miles each?

   Moscow, Russia

3. Which subways could be evenly broken into sections of 5 miles each?

   Paris, France, and Tokyo, Japan

Circle the letter of the correct answer.

4. Which subway's length is divisible by 4 miles?
   - **A** New York, United States
   - **B** Paris, France
   - **C** Tokyo, Japan
   - **(D)** Moscow, Russia

5. Which subway's length is not a prime number, but is also not divisible by 2, 3, 4, 5, 6, or 9?
   - **F** Mexico City, Mexico
   - **(G)** New York, United States
   - **H** Seoul, South Korea
   - **J** Paris, France

6. The subway in Hong Kong, China, has a length that is a prime number of miles. Which of the following is its length?
   - **A** 260 miles
   - **B** 268 miles
   - **(C)** 269 miles
   - **D** 265 miles

7. The subway in St. Petersburg, Russia, has a length that is divisible by 3 miles. Which of the following is its length?
   - **(F)** 57 miles
   - **G** 56 miles
   - **H** 55 miles
   - **J** 58 miles

LESSON **4-1**
Reading Strategies
Use a Graphic Aid

1	2	3	4	5	6	7	8	9	10
11	12	13	14	15	16	17	18	19	20
21	22	23	24	25	26	27	28	29	30
31	32	33	34	35	36	37	38	39	40
41	42	43	44	45	46	47	48	49	50
51	52	53	54	55	56	57	58	59	60
61	62	63	64	65	66	67	68	69	70
71	72	73	74	75	76	77	78	79	80
81	82	83	84	85	86	87	88	89	90
91	92	93	94	95	96	97	98	99	100

Circle all the multiples of 2 on the hundred chart.

1. Write the first number you circled in the second row.

   12

2. Divide 12 by 2. What is the answer?

   6

Divisible means **"can be divided by."** A number is divisible by another number if the quotient is a whole number with no remainder. 12 can be evenly divided by 2. All multiples of 2 are divisible by 2.

Circle the multiples of 3, 5, and 10 on the chart in three different colors.

3. Is each number that is divisible by 10 also divisible by 5? Why or why not?

   Yes; all numbers that end in 0 are circled for both 10 and 5.

4. Is every number that is divisible by 5 also divisible by 10? Why or why not?

   No; numbers that end in 5 are not divisible by 10.

5. Is every number that is divisible by 5 also divisible by 3? Why or why not?

   No; different multiples are circled for 3 and 5.

LESSON 4-1
Puzzles, Twisters & Teasers
Monkey's Web

Monkeys swinging among the trees created webs of tangled vines. There is an inner web and an outer web. It's your job to connect them so that the monkeys can run between them. Two points in the web can be connected by a straight line only if the number in the outer ring is evenly divisible by the number in the inner ring. Two points cannot be connected if the straight line would cross through the inner ring.

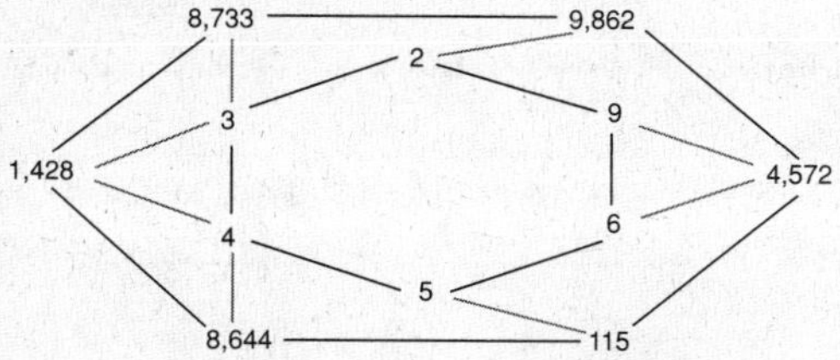

9,862 was divisible by 2 = P

8,733 was divisible by 3 = I

8,644 was divisible by 4 = S

4,572 was divisible by 9, 6 = M

115 was divisible by 5 = T

To solve the riddle, fill in the letters that match your answers below.

What happened when Monkey 3 married Monkey 5?

They became P R I M E M A T E S
(2 _ 3 9 _ 6 _ 5 _ 4)

LESSON 4-2
Practice A
Factors and Prime Factorization

List all of the factors of each number.

1. 6 — 1; 2; 3; 6
2. 9 — 1; 3; 9
3. 10 — 1; 2; 5; 10
4. 12 — 1; 2; 3; 4; 6; 12
5. 21 — 1; 3; 7; 21
6. 18 — 1; 2; 3; 6; 9; 18
7. 16 — 1; 2; 4; 8; 16
8. 25 — 1; 5; 25
9. 31 — 1; 31

Write the prime factorization of each number.

10. 9 — 3^2
11. 25 — 5^2
12. 8 — 2^3
13. 14 — $2 \cdot 7$
14. 12 — $2^2 \cdot 3$
15. 15 — $3 \cdot 5$
16. 5 — 5
17. 20 — $2^2 \cdot 5$
18. 26 — $2 \cdot 13$

19. There are 12 chairs in the meeting hall and an odd number of tables. What are all the possible ways the chairs could be arranged so that each table has the same number of chairs?

12 chairs at 1 table or 4 chairs at 3 tables

20. What are two different ways that 9 can be written as a product of two numbers?

Possible responses: $9 \cdot 1$; $1 \cdot 9$; $3 \cdot 3$

LESSON 4-2
Practice B
Factors and Prime Factorization

List all of the factors of each number.

1. 15 — 1; 3; 5; 15
2. 24 — 1; 2; 3; 4; 6; 8; 12; 24
3. 33 — 1; 3; 11; 33
4. 72 — 1; 2; 3; 4; 6; 8; 9; 12; 18; 24; 36; 72
5. 48 — 1; 2; 3; 4; 6; 8; 12; 16; 24; 48
6. 95 — 1; 5; 19; 95
7. 66 — 1; 2; 3; 6; 11; 22; 33; 66
8. 87 — 1; 3; 29; 87
9. 36 — 1; 2; 3; 4; 6; 9; 12; 18; 36

Write the prime factorization of each number.

10. 44 — $2^2 \cdot 11$
11. 56 — $2^3 \cdot 7$
12. 42 — $2 \cdot 3 \cdot 7$
13. 39 — $3 \cdot 13$
14. 36 — $2^2 \cdot 3^2$
15. 125 — 5^3
16. 85 — $5 \cdot 17$
17. 100 — $2^2 \cdot 5^2$
18. 32 — 2^5

19. James has an assigned seat for his flight to Denver. The seats on the plane are numbered 1–49. James's seat number is an odd number greater than 10 that is factor of 100. What is his seat number for the flight?

25

20. Linda writes the prime factorization of 40 as $2 \cdot 2 \cdot 2 \cdot 5$ on the board. Phil writes the prime factorization of 40 as $2^3 \cdot 5$. Who is correct?

They both are.

LESSON 4-2
Practice C
Factors and Prime Factorization

List all of the factors of each number.

1. 92 — 1; 2; 4; 23; 46; 92
2. 356 — 1; 2; 4; 89; 178; 356
3. 180 — 1; 2; 3; 4; 5; 6; 9; 10; 12; 15; 18; 20; 30; 36; 45; 60; 90; 180
4. 550 — 1; 2; 5; 10; 11; 22; 25; 50; 55; 110; 275; 550

Write the prime factorization of each number.

5. 225 — $3^2 \cdot 5^2$
6. 333 — $3^2 \cdot 37$
7. 124 — $2^2 \cdot 31$
8. 216 — $2^3 \cdot 3^3$
9. 423 — $3^2 \cdot 47$
10. 810 — $2 \cdot 3^4 \cdot 5$

Write each number as a product in two different ways. Possible answers given.

11. 81 — $1 \cdot 81$; $9 \cdot 9$; $3 \cdot 27$
12. 117 — $1 \cdot 117$; $3 \cdot 39$; $9 \cdot 13$
13. 375 — $1 \cdot 375$; $3 \cdot 125$; $5 \cdot 75$; $15 \cdot 25$

14. You and your friend are going to split the houses in your neighborhood for newspaper delivery. Which would you prefer, the odd-numbered houses or the prime-numbered houses? Explain.

If students want more houses, they should choose odd numbers. If they want fewer houses, they should choose prime numbers, because there are fewer prime numbers than odd numbers.

15. Selma cannot remember her locker number! She knows that her locker number is prime, and that it is a factor of 435. What are all of Selma's possible locker numbers?

3, 5, or 29

LESSON **4-2**

Reteach
Factors and Prime Factorization

Factors of a product are the numbers that are multiplied to find that product. A factor is also a whole number that divides the product with no remainder.

To find all of the factors of 24, make a list of multiplication facts.

$1 \cdot 24 = 24$
$2 \cdot 12 = 24$
$3 \cdot 8 = 24$
$4 \cdot 6 = 24$

The factors of 24 are 1, 2, 3, 4, 6, 8, 12, and 24.

Write multiplication facts to find the factors of each number.

1. 20
   $1 \bullet 20 = 20; 2 \bullet 10 = 20;$
   $4 \bullet 5 = 20$

2. 16
   $1 \bullet 16 = 16; 2 \bullet 8 = 16;$
   $4 \bullet 4 = 16$

3. 35
   $1 \bullet 35 = 35; 5 \bullet 7 = 35$

4. 31
   $1 \bullet 31 = 31$

A number written as the product of prime factors is called the prime factorization of the number.

To write the prime factorization of 24, first write it as product of 2 numbers. Then rewrite each factor as the product of 2 numbers until all of the factors are prime numbers.

$24 = \mathbf{4} \cdot 6$ (Write 24 as the product of 2 numbers.)
$= 2 \cdot 2 \cdot \mathbf{6}$ (Rewrite 4 as the product of 2 prime numbers.)
$= 2 \cdot 2 \cdot 2 \cdot 3$ (Rewrite 6 as the product of 2 prime numbers.)

So, the prime factorization of 24 is $2 \cdot 2 \cdot 2 \cdot 3$ or $2^3 \cdot 3$.

Find the prime factorization of each number.

5. 28 $2^2 \bullet 7$
6. 45 $3^2 \bullet 5$
7. 50 $2 \bullet 5^2$
8. 72 $2^3 \bullet 3^2$

LESSON **4-2**

Challenge
Prime Shades

For each given number, shade one box in each row of the table to show a prime factor. Then use your shaded boxes to write each number's prime factorization.

1. 12

7	2
3	5
2	11

Prime factorization: $2^2 \bullet 3$

2. 70

5	3
2	11
3	7

Prime factorization: $2 \bullet 5 \bullet 7$

3. 63

3	5
2	3
11	7

Prime factorization: $3^2 \bullet 7$

4. 150

13	5
3	7
2	11
17	5

Prime factorization: $2 \bullet 3 \bullet 5^2$

5. 84

11	7
5	2
13	2
3	17

Prime factorization: $2^2 \bullet 3 \bullet 7$

6. 260

5	11
3	2
17	13
2	7

Prime factorization: $2^2 \bullet 5 \bullet 13$

7. 80

5	7
17	2
2	13
11	2
2	3

Prime factorization: $2^4 \bullet 5$

8. 1,750

17	5
13	7
31	5
5	3
2	11

Prime factorization: $2 \bullet 5^3 \bullet 7$

9. 3,234

5	3
17	7
31	11
2	13
7	23

Prime factorization: $2 \bullet 3 \bullet 7^2 \bullet 11$

LESSON **4-2**

Problem Solving
Factors and Prime Factorization

Write the correct answer.

1. The area of a rectangle is the product of its length and width. If a rectangular board has an area of 30 square feet, what are the possible measurements of its length and width?
   1, 2, 3, 5, 6, 10, 15, or 30 feet

2. The first-floor apartments in Jenna's building are numbered 100 to 110. How many apartments on that floor are a prime number? What are those apartment numbers?
   4 apartments; 101, 103, 107, and 109

3. A Russian mathematician named Christian Goldbach came up with a theory that every even number greater than 4 can be written as the sum of two odd primes. Test Goldbach's theory with the numbers 6 and 50. **Possible answers:**
   6 = 3 + 3;
   50 = 19 + 31

4. Mr. Samuels has 24 students in his math class. He wants to divide the students into equal groups, and he wants the number of students in each group to be prime. What are his choices for group sizes? How many groups can he make?
   12 groups of 2 students each or 8 groups of 3 students each

Circle the letter of the correct answer.

5. Why is 2 the only even prime number?
   A It is the smallest prime number.
   (B) All other even numbers are divisible by 2.
   C It only has 1 and 2 as factors.
   D All odd numbers are prime.

6. What prime numbers are factors of both 60 and 105?
   F 2 and 3
   G 2 and 5
   (H) 3 and 5
   J 5 and 7

7. If a composite number has the first five prime numbers as factors, what is the smallest number it could be? Write that number's prime factorization.
   A 30
   B 210
   (C) 2,310
   D 30,030

8. Tim's younger brother, Bryant, just had a birthday. Bryant's age only has one factor, and is not a prime number. How old is Bryant?
   F 10 years old
   G 7 years old
   H 3 years old
   (J) 1 year old

LESSON **4-2**

Reading Strategies
Use an Organizer

A **prime number** has only two factors, 1 and the number itself.
2 ⟶ $2 \cdot 1 = 2$ The factors are 1 and 2.

A **composite number** has more than 2 factors.
6 ⟶ $1 \cdot 6 = 6$ and $2 \cdot 3 = 6$ The factors are 1, 2, 3, and 6.

When a composite number is written as the product of prime numbers, it is called a **prime factorization,** such as $12 = 2 \times 2 \times 3$.

Drawing a **factor tree** will help you find and organize the prime factors of a number such as 42.

42 ⟵ Write the number.
6 (7) ⟵ Write any pair of factors. Circle 7 because it is a prime number.
(2) (3) ⟵ Continue until all factors are prime.
$2 \times 3 \times 7$ ⟵ Write the prime factors from least to greatest.

Answer each question.

1. Write a pair of factors you could use to find the prime factorization of 30.
   5 and 6, 3 and 10, or 2 and 15

2. Use the factors in the answer above to write the prime factorization of 30.
   $2 \times 3 \times 5 = 30$

3. Make a factor tree for 36. What pair of factors did you start with?
   Sample answers: 9 and 4 or 6 and 6

4. What is the prime factorization of 36?
   $2 \times 2 \times 3 \times 3 = 36$

5. Make a factor tree for 16. What pair of factors did you start with?
   Sample answers: 4 and 4 or 8 and 2

6. What is the prime factorization for 16?
   $2 \times 2 \times 2 \times 2 = 16$

LESSON 4-2

Puzzles, Twisters & Teasers

Get Cooking

For each number in the left-hand column, there is matching prime factorization in the right-hand column.

Connect the numbers on the left to the factorizations on the right for the correct match. Each line will pass through a letter in the center.

Fill in the letters below to solve the riddle.

1.	56	I	3 × 3 × 5 × 5
2.	225	M	2 × 3 × 7 × 7
		T	
3.	61	P	2 × 2 × 2 × 7
		A	
4.	294	E	3 × 3 × 3 × 37
		S	
5.	999	T	prime
		D	
6.	77	B	2 × 3 × 3 × 3 × 5
		R	
7.	41		7 × 11

What did the whole number serve at his family barbecue?

P	R	I	M	E
4	6	2	1	3

R	I	B	S
6	2	7	5

LESSON 4-3

Practice A

Greatest Common Factor

Find the GCF of each set of numbers.

1. 6 and 9 — 3
2. 4 and 8 — 4
3. 8 and 12 — 4
4. 6 and 15 — 3
5. 10 and 15 — 5
6. 9 and 12 — 3
7. 4 and 10 — 2
8. 5 and 20 — 5
9. 7 and 14 — 7
10. 8 and 11 — 1
11. 2 and 12 — 2
12. 9 and 21 — 3

Circle the letter of the correct answer.

13. For which set of numbers is 6 the GCF?
    - A 2, 3, and 6
    - B 3, 6, and 12
    - (C) 12, 18, and 24
    - D 1, 6, and 12

14. For which set of numbers is 4 the GCF?
    - F 1, 4, and 8
    - G 2, 4, and 16
    - H 1, 2, and 4
    - (J) 8, 12, and 16

15. Bonny has 24 wood beads and 30 glass beads. She wants each necklace she makes to have the same number of wood beads and the same number of glass beads. What is the greatest number of necklaces she can make if all the beads are used?

    6 necklaces

16. Mike is setting up fish tanks at the pet store. He has 6 angel fish, 12 tiger barbs, and 15 guppies. If he wants to have the same number of each kind of fish in every tank, what is the greatest number of tanks he can set up?

    3 fish tanks

LESSON 4-3

Practice B

Greatest Common Factor

Find the GCF of each set of numbers.

1. 12 and 15 — 3
2. 18 and 24 — 6
3. 15 and 25 — 5
4. 16 and 24 — 8
5. 36 and 45 — 9
6. 24 and 54 — 6
7. 48 and 64 — 16
8. 27 and 72 — 9
9. 55 and 77 — 11
10. 16, 28, and 48 — 4
11. 15, 35, and 95 — 5
12. 20, 30, and 80 — 10
13. 18, 36, and 54 — 18
14. 27, 36, and 45 — 9
15. 21, 49, and 63 — 7
16. 25, 35, and 45 — 5
17. 28, 42, and 63 — 7
18. 25, 75, and 115 — 5

19. Mr. Thompson's sixth-grade class is competing in the school field day. There are 16 boys and 12 girls in his class. He divided the class into the greatest number of teams possible with the same number of boys on each team and the same number of girls on each team. How many teams were made if each person was on a team? How many girls were on each team? How many boys?

    4 teams with 3 girls and 4 boys on each team

20. Barbara is making candy bags for her birthday party. She has 24 lollipops, 12 candy bars, and 42 pieces of gum. She wants each bag to have the same number of each kind of candy. What is the greatest number of bags she can make if all the candy is used? How many pieces of each kind of candy will be in each bag?

    6 bags with 4 lollipops, 2 candy bars, and 7 pieces of gum in each bag

LESSON 4-3

Practice C

Greatest Common Factor

Find the GCF of each set of numbers.

1. 48 and 64 — 16
2. 72 and 81 — 9
3. 54 and 66 — 6
4. 56 and 80 — 8
5. 36 and 48 — 12
6. 32 and 232 — 8
7. 20, 44, and 88 — 4
8. 72, 96, and 84 — 12
9. 54, 60, and 78 — 6
10. 49, 84, and 91 — 7
11. 150, 200, and 300 — 50
12. 88, 96, and 120 — 8
13. 81, 108, and 117 — 9
14. $3^3 \cdot 4$ and $3^2 \cdot 4^2$ — 36
15. $2^2 \cdot 3 \cdot 5$ and $2^3 \cdot 3^3$ — 12
16. 2, 8, 16, and 40 — 2
17. 12, 15, 20, and 30 — 1
18. 5, 10, 15, and 25 — 5

19. Mrs. Hill is preparing for group art projects. She has 52 pipe cleaners, 78 pieces of colored paper, 91 cotton balls, and 117 strings of yarn. Each group will get the same number of each kind of art supply. What is the greatest number of groups that Ms. Hill can make if all the supplies are used? How many of each item will each group get?

    13 groups with 4 pipe cleaners, 6 pieces of colored paper, 7 cotton balls, and 9 strings of yarn for each group

20. Tony is making flower arrangements. He is using three different kinds of flowers—roses, daisies, and tulips. Each arrangement will have the same number of each kind of flower. He figures that if he uses all the flowers, the greatest number of arrangements he can make is 7. He has 42 flowers in all, with more daisies than tulips, and more tulips than roses. How many of each kind of flower does Tony have?

    He has 21 daisies, 14 tulips, and 7 roses.

LESSON 4-3
Reteach
Greatest Common Factor

The greatest common factor, or GCF, is the largest number that is the factor of any set of at least two numbers.

You can use prime factorization to find the GCF of two or more numbers.

To find the GCF of 18 and 24, first write the prime factorization of each number. Then identify the common prime factors.

$18 = \mathbf{2} \cdot 3 \cdot \mathbf{3}$
$24 = \mathbf{2} \cdot 2 \cdot 2 \cdot \mathbf{3}$

Next, find the product of the common prime factors.
$2 \cdot 3 = 6$

The GCF of 18 and 24 is 6.

Find the GCF of each set of numbers.

1. 32 and 48
$32 = 2^5$
$48 = 2^4 \cdot 3$
16

2. 45 and 81
$45 = 3^2 \cdot 5$
$81 = 3^4$
9

3. 18 and 36
$18 = 2 \cdot 3^2$
$36 = 2^3 \cdot 3^2$
18

4. 14 and 35
$14 = 2 \cdot 7$
$35 = 5 \cdot 7$
7

5. 42 and 72
$42 = 2 \cdot 3 \cdot 7$
$72 = 2^3 \cdot 3^2$
6

6. 56 and 64
$56 = 2^2 \cdot 7$
$64 = 2^6$
8

7. 28, 56, and 84
$28 = 2^2 \cdot 7$
$56 = 2^3 \cdot 7$
$84 = 2^2 \cdot 3 \cdot 7$
28

8. 30, 45, and 75
$30 = 2 \cdot 3 \cdot 5$
$45 = 3^2 \cdot 5$
$75 = 3 \cdot 5^2$
15

9. 36, 45, and 54
$36 = 2^2 \cdot 3^2$
$45 = 3^2 \cdot 5$
$54 = 2 \cdot 3^3$
9

LESSON 4-3
Challenge
The Greatest Common Flower

A florist made these flower arrangements for a wedding. He used every flower in each crate he had to create the greatest number of arrangements possible. Study the flowers the florist had in each crate on the left. Below each, write the number of arrangements the florist made with those flowers. Then draw a line to the correct arrangement on the right that the florist created with those flowers.

1. 6 18 24
Number of arrangements: 6

2. 12 16 20
Number of arrangements: 4

3. 15 30 35
Number of arrangements: 5

4. 21 42 56
Number of arrangements: 7

5. 9 15 18
Number of arrangements: 3

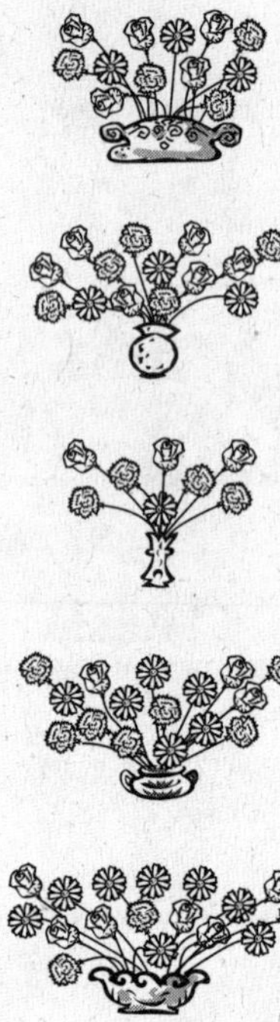

LESSON 4-3
Problem Solving
Greatest Common Factor

Write the correct answer.

1. Carolyn has 24 bottles of shampoo, 36 tubes of hand lotion, and 60 bars of lavender soap to make gift baskets. She wants to have the same number of each item in every basket. What is the greatest number of baskets she can make without having any of the items left over?

12 baskets

2. There are 40 girls and 32 boys who want to participate in the relay race. If each team must have the same number of girls and boys, what is the greatest number of teams that can race? How many boys and girls will be on each team?

8 teams with 5 girls and 4 boys each

3. Ming has 15 quarters, 30 dimes, and 48 nickels. He wants to group his money so that each group has the same number of each coin. What is the greatest number of groups he can make? How many of each coin will be in each group? How much money will each group be worth?

3 groups with 5 quarters, 10 dimes, and 16 nickels each; $3.05

4. A gardener has 27 tulip bulbs, 45 tomato plants, 108 rose bushes, and 126 herb seedlings to plant in the city garden. He wants each row of the garden to have the same number of each kind of plant. What is the greatest number of rows that the gardener can make if he uses all the plants?

9 rows

Circle the letter of the correct answer.

5. Kim packed 6 boxes with identical supplies. It was the greatest number she could pack and use all the supplies. Which of these is her supply list?

A 24 pencils, 36 pens, 10 rulers
B 12 rulers, 30 pencils, 45 pens
(C) 42 pencils, 18 rulers, 72 pens
D 60 pens, 54 pencils, 32 rulers

6. The sum of three numbers is 60. Their greatest common factor is 4. Which of the following lists shows those three numbers?

F 4, 16, 36
(G) 8, 20, 32
H 14, 16, 30
J 10, 18, 32

LESSON 4-3
Reading Strategies
Follow a Procedure

Numbers you multiply together are called **factors.** When you multiply factors together, the answer is called the **product.**

Answer each question.

1. Is 2 a factor of 8? Why or why not?

Yes, because you can multiply 2 times 4 to get 8

2. Is 3 a factor of 8? Why or why not?

No, because you cannot multiply 3 by any number to get 8

The word **common** means "shared." All the students in a class who have blue eyes have eye color in common.
Whole numbers may share the same factors. Shared factors are called **common factors.**

These are the factor pairs for 12 and 18.

12:	$1 \cdot 12$	$2 \cdot 6$	$3 \cdot 4$
18:	$1 \cdot 18$	$2 \cdot 9$	$3 \cdot 6$

Answer the following questions.

3. List all of the factors for 12.

1, 2, 3, 4, 6, 12

4. List all of the factors for 18.

1, 2, 3, 6, 9, 18

5. Find the factors that 12 and 18 share in common.

1, 2, 3, 6

The largest factor that two or more whole numbers share is called the **greatest common factor.**

6. What is the greatest common factor for 12 and 18?

6

7. List the steps you use to find the greatest common factor of two numbers.

Sample answer: list the factors of 12 and 18 and find the largest factor they share.

LESSON 4-3

Puzzles, Twisters & Teasers

Magical Factor Cruise

This is the game of Magical Factor Island. You are currently on a cruise ship and must get to this island in order to win a magical prize.

You must decide if the statement in each box is true or false. If it is false black out the box. Once you have all the false statements blacked out you will be able to cruise to the magical island.

GCF of 12 and 8 = 4	GCF of 5 and 12 = 5	GCF of 42 and 28 = 12	GCF of 63 and 18 = 18	GCF of 96 and 64 = 30
GCF of 6 and 24 = 6	GCF of 8 and 20 = 4	GCF of 12 and 48 = 12	GCF of 50 and 125 = 15	GCF of 70 and 20 = 5
GCF of 15 and 27 = 9	GCF of 12 and 72 = 4	GCF of 10 and 15 = 5	GCF of 48 and 36 = 4	GCF of 18 and 24 = 4
GCF of 32 and 48 = 2	GCF of 24 and 66 = 2	GCF of 24 and 32 = 8	GCF of 15 and 27 = 3	GCF of 6 and 18 = 6
GCF of 16 and 33 = 4	GCF of 6 and 18 = 3	GCF of 12 and 24 = 6	GCF of 24 and 56 = 2	GCF of 20 and 40 = 20
GCF of 9 and 45 = 5	GCF of 18 and 30 = 3	GCF of 75 and 90 = 10	GCF of 40 and 100 = 40	GCF of 16 and 40 = 8

LESSON 4-4

Practice A

Decimals and Fractions

Write each decimal as a fraction or mixed number.

1. 0.5 — $\frac{5}{10}$ or $\frac{1}{2}$
2. 0.25 — $\frac{25}{100}$ or $\frac{1}{4}$
3. 0.75 — $\frac{75}{100}$ or $\frac{3}{4}$
4. 0.4 — $\frac{4}{10}$ or $\frac{2}{5}$
5. 0.8 — $\frac{8}{10}$ or $\frac{4}{5}$
6. 1.2 — $1\frac{2}{10}$ or $1\frac{1}{5}$

Write each fraction or mixed number as a decimal.

7. $\frac{3}{10}$ — 0.3
8. $\frac{1}{2}$ — 0.5
9. $\frac{1}{4}$ — 0.25
10. $1\frac{3}{4}$ — 1.75
11. $\frac{3}{5}$ — 0.6
12. $1\frac{2}{5}$ — 1.4

Circle the letter of the best answer.

13. Which of the following sets is written in order from least to greatest?
   - A 0.5, $\frac{1}{4}$, 0.75
   - B 0.4, $\frac{7}{10}$, 0.6
   - (C) $\frac{1}{4}$, 0.5, 0.75
   - D $\frac{7}{10}$, 0.4, 0.6

14. Which of the following sets is written in order from greatest to least?
   - F $\frac{1}{3}$, $1\frac{1}{2}$, $1\frac{3}{4}$
   - (G) $\frac{2}{5}$, $0.\overline{3}$, 0.3
   - H $1\frac{1}{2}$, $1\frac{3}{4}$, $1\frac{1}{3}$
   - J 0.3, $0.\overline{3}$, $\frac{2}{5}$

15. At Franklin Elementary School, $\frac{2}{3}$ of all the students attended the chorus recital on Thursday. On Friday, $\frac{3}{4}$ of all the students attended the basketball game. Which event had the highest attendance?

   the basketball game

16. Each of the Girl Scout troops was given the same number of cookies to sell. Tina's troop sold 0.25 of its cookies. Chantall's Girl Scout troop sold half of its cookies. Which troop sold the most cookies?

   Chantall's troop

LESSON 4-4

Practice B

Decimals and Fractions

Write each decimal as a fraction or mixed number.

1. 0.23 — $\frac{23}{100}$
2. 0.1 — $\frac{1}{10}$
3. 3.25 — $3\frac{25}{100}$ or $3\frac{1}{4}$
4. $1.\overline{3}$ — $1\frac{1}{3}$
5. 5.5 — $5\frac{5}{10}$ or $5\frac{1}{2}$
6. 3.7 — $3\frac{7}{10}$

Write each fraction or mixed number as a decimal.

7. $\frac{4}{5}$ — 0.8
8. $\frac{1}{9}$ — $0.\overline{1}$
9. $1\frac{2}{3}$ — $1.\overline{6}$
10. $3\frac{3}{5}$ — 3.6
11. $2\frac{1}{3}$ — $2.\overline{3}$
12. $\frac{8}{9}$ — $0.\overline{8}$

Order the fractions and decimals from least to greatest.

13. $\frac{1}{4}$, 0.7, $\frac{3}{5}$ — $\frac{1}{4}$, $\frac{3}{5}$, 0.7
14. 0.25, $\frac{1}{8}$, 0.3 — $\frac{1}{8}$, 0.25, 0.3
15. $\frac{9}{10}$, 0.49, $\frac{1}{2}$ — 0.49, $\frac{1}{2}$, $\frac{9}{10}$

Order the fractions and decimals from greatest to least.

16. 0.13, $\frac{1}{10}$, 0.9 — 0.9, 0.13, $\frac{1}{10}$
17. $\frac{2}{5}$, 0.7, $\frac{2}{3}$ — 0.7, $\frac{2}{3}$, $\frac{2}{5}$
18. 0.65, $\frac{4}{5}$, $\frac{3}{4}$ — $\frac{4}{5}$, $\frac{3}{4}$, 0.65

19. Derrick has a dollar bill and three dimes, Jane has a dollar bill and one quarter, and Kelly has a dollar bill and ten nickels. Who has the most money? the least?

   Kelly has the most, and Jane has the least.

20. It rained three and one half inches in April. In May it rained $3\frac{3}{4}$ inches, and in June it rained 3.6 inches. Write the months in order from the greatest to the least amount of rain.

   May, June, April

LESSON 4-4

Practice C

Decimals and Fractions

Write each decimal as a fraction or mixed number.

1. 0.97 — $\frac{97}{100}$
2. 2.03 — $2\frac{3}{100}$
3. 56.1 — $56\frac{1}{10}$
4. $6.\overline{6}$ — $6\frac{2}{3}$
5. 7.53 — $7\frac{53}{100}$
6. 12.009 — $12\frac{9}{1,000}$

Write each fraction or mixed number as a decimal.

7. $1\frac{8}{9}$ — $1.\overline{8}$
8. $\frac{1}{12}$ — $0.08\overline{3}$
9. $2\frac{7}{100}$ — 2.07
10. $\frac{31}{500}$ — 0.062
11. $\frac{17}{20}$ — 0.85
12. $\frac{5}{11}$ — $0.\overline{45}$

Order the fractions and decimals from least to greatest.

13. 0.83, $\frac{7}{8}$, $\frac{4}{5}$ — $\frac{4}{5}$, 0.83, $\frac{7}{8}$
14. $\frac{9}{11}$, 0.9, $\frac{5}{6}$ — $\frac{9}{11}$, $\frac{5}{6}$, 0.9
15. $4\frac{2}{3}$, $4.\overline{2}$, $4\frac{3}{11}$ — $4.\overline{2}$, $4\frac{3}{11}$, $4\frac{2}{3}$

Order the fractions and decimals from greatest to least.

16. $\frac{3}{10}$, $\frac{27}{100}$, $\frac{1}{3}$ — $\frac{1}{3}$, $\frac{3}{10}$, $\frac{27}{100}$
17. $\frac{8}{12}$, $\frac{3}{4}$, 0.71 — $\frac{3}{4}$, 0.71, $\frac{8}{12}$
18. $\frac{19}{20}$, $\frac{97}{100}$, 0.99 — 0.99, $\frac{97}{100}$, $\frac{19}{20}$

19. Shawn has 3.65 cups of sugar and 1.9 cups of flour. He needs $3\frac{2}{3}$ cups of sugar and $1\frac{7}{8}$ cups of flour to make cookies. Does he have enough of each ingredient?

   He has enough flour, but not enough sugar.

20. This week, Katie ran $12\frac{11}{12}$ miles, Sandra ran 12.91 miles, and Tameeka ran $12\frac{9}{10}$ miles. Who ran the most miles this week? the least?

   Katie ran the most, and Tameeka ran the least.

LESSON **4-4**

Reteach

Decimals and Fractions

You can write decimals as fractions or mixed numbers. A place value chart will help you read the decimal. Remember the decimal point is read as the word "and."

To write 0.47 as a fraction, first think about the decimal in words.

Ones	Tenths	Hundredths	Thousandths	Ten Thousandths
0.	4	7		

0.47 is read "forty-seven hundredths." The place value of the decimal tells you the denominator is 100.

$0.47 = \frac{47}{100}$

To write 8.3 as a mixed number, first think about the decimal in words.

Ones	Tenths	Hundredths	Thousandths	Ten Thousandths
8.	3			

8.3 is read "eight and three tenths." The place value of the decimal tells you the denominator is 10. The decimal point is read as the word "and."

$8.3 = 8\frac{3}{10}$

Write each decimal as a fraction or mixed number.

1. 0.61 $\frac{61}{100}$

2. 3.43 $3\frac{43}{100}$

3. 0.009 $\frac{9}{1,000}$

4. 4.7 $4\frac{7}{10}$

5. 1.5 $1\frac{5}{10}$ or $1\frac{1}{2}$

6. 0.13 $\frac{13}{100}$

7. 5.002 $5\frac{2}{1,000}$ or $5\frac{1}{500}$

8. 0.021 $\frac{21}{1,000}$

LESSON **4-4**

Reteach

Decimals and Fractions (continued)

Fractions and mixed numbers can be written as decimals.

To write $\frac{1}{4}$ as a decimal, first think about the expression in words.

$\frac{1}{4}$ means "1 divided by 4."

Then do the division.

$$\begin{array}{r} 0.25 \\ 4\overline{)1.00} \\ -8 \\ \hline 20 \\ -20 \\ \hline 0 \end{array}$$

0.25 is a terminating decimal because it has an end.

$\frac{1}{4} = 0.25$

A number that contains a whole number and a fraction is called a mixed number. $2\frac{1}{3}$ is an example of a mixed number. To write $2\frac{1}{3}$ as a decimal, first think about the expression in words. $2\frac{1}{3}$ means "2 and 1 divided by 3." Keep 2 as the whole number.

Then do the division.

$$\begin{array}{r} 0.33 \\ 3\overline{)1.00} \\ -9 \\ \hline 10 \\ -09 \\ \hline 1 \end{array}$$

$1 \div 3 = 0.33\ldots$, or $0.\overline{3}$. $0.\overline{3}$ is a repeating decimal because it does not end.

So, $2\frac{1}{3} = 2.\overline{3}$.

Write each fraction or mixed number as a decimal.

9. $\frac{3}{5}$ 0.6

10. $3\frac{3}{4}$ 3.75

11. $\frac{2}{3}$ $0.\overline{6}$

12. $1\frac{2}{9}$ $1.\overline{2}$

13. $\frac{1}{6}$ $0.1\overline{6}$

14. $2\frac{1}{8}$ 2.125

15. $\frac{5}{6}$ $0.8\overline{3}$

16. $8\frac{1}{9}$ $8.\overline{1}$

LESSON **4-4**

Challenge

Fractions of Pizza

Write the fractions as decimals. Then slice and shade pieces of each pizza to represent each fraction. Finally, compare the pizzas in each row by writing <, >, or =.

1. $\frac{3}{4}$ 0.75 $\frac{4}{5}$ 0.8

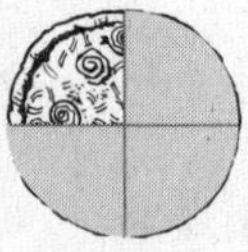
<
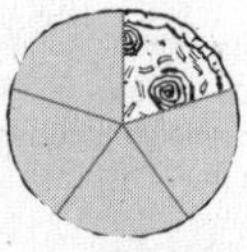

2. $\frac{3}{6}$ 0.5 $\frac{1}{2}$ 0.5

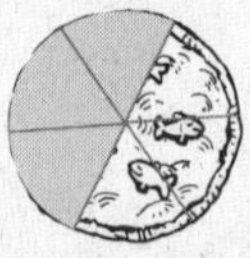
=

3. $\frac{2}{3}$ $0.\overline{6}$ $\frac{3}{5}$ 0.6

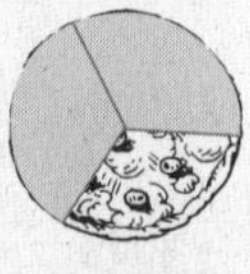
>
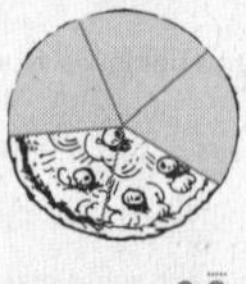

4. $\frac{1}{5}$ 0.2 $\frac{1}{3}$ $0.\overline{3}$

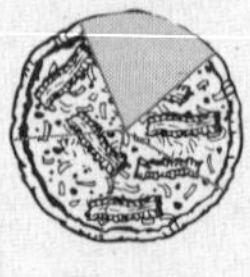
<
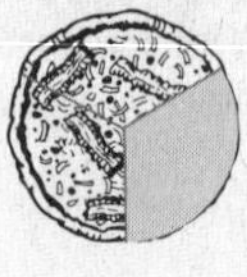

LESSON **4-4**

Problem Solving

Decimals and Fractions

Electricity is measured in amperes, or the rate electrical currents flow. A high ampere measurement means that a lot of electricity is being used. The table below shows the average amount of electricity some household appliances use per hour. Use the table to answer the questions.

Electricity Use in the Home

Appliance	Amps per Hour
Blender	$2\frac{1}{2}$
Coffeemaker	$6\frac{2}{3}$
Computer and printer	$1\frac{5}{6}$
Microwave oven	$12\frac{1}{2}$
Popcorn popper	$2\frac{1}{12}$
25-inch television	$1\frac{1}{4}$
VCR	$\frac{1}{3}$

1. How much electricity does an average 25-inch television use each hour? Write your answer as a decimal.

   1.25 amperes

2. Which appliance uses an average of 2.5 amps per hour?

   blender

3. Which appliance uses the most electricity per hour? Write its ampere measurement as a decimal.

   microwave oven;
   12.5 amperes

Circle the letter of the correct answer.

4. How much electricity do most computers and printers use in an hour?

   A 1.38 amperes
   B 1.8 amperes
   (C) $1.8\overline{3}$ amperes
   D 1.88 amperes

5. Which of the appliances has an hourly ampere measurement that is a repeating decimal?

   F blender
   (G) coffee maker
   H microwave oven
   J 25-inch television

6. In most years, 39.7 percent of the world's energy comes from burning oil. What is this percent written as a fraction?

   A $\frac{39}{7}$ percent
   B $39\frac{1}{7}$ percent
   C $3\frac{9}{7}$ percent
   (D) $39\frac{7}{10}$ percent

7. The United States produces about 13.2 percent of the world's hydroelectric power. What fraction of hydroelectric power does the United States produce?

   (F) $13\frac{1}{5}$ percent
   G $\frac{13}{2}$ percent
   H $1\frac{3}{2}$ percent
   J $13\frac{1}{2}$ percent

LESSON **4-4**

Reading Strategies

Focus on Vocabulary

The word *repeating* means something happening over and over. A church bell that chimes every hour is an example of a repeating sound.

The word *terminating* means something that ends. The sixth grade class party will be terminating at 10:00 P.M.

Fractions can be written as either **terminating** or **repeating decimals.**

In a repeating decimal, sometimes one digit repeats and sometimes more than one digit repeats.

$\frac{1}{3}$ → means 1 ÷ 3

$$\begin{array}{r} 0.333 \\ 3\overline{)1.000} \\ -9 \\ \hline 10 \\ -9 \\ \hline 10 \\ -9 \\ \hline 1 \end{array}$$

$\frac{1}{3} = 0.\overline{3}$

The bar over 3 means the 3 keeps repeating.

$\frac{1}{4}$ → means 1 ÷ 4

$$\begin{array}{r} 0.25 \\ 4\overline{)1.00} \\ -8 \\ \hline 20 \\ 20 \\ \hline 0 \end{array}$$

$\frac{1}{4} = 0.25$

0.25 is a decimal that terminates.

Write "terminating decimal" or "repeating decimal" for each. If the decimal repeats, write which digit or digits repeat.

1. 0.166666 repeating decimal; 6
2. 0.125 terminating decimal
3. 0.090909 repeating decimal; 09
4. 0.2222 repeating decimal; 2
5. 0.8 terminating decimal
6. 0.555 repeating decimal; 5
7. 0.5 terminating decimal
8. 0.99 repeating decimal; 9

LESSON **4-4**

Puzzles, Twisters & Teasers

How High Can You Climb?

For each decimal, fraction and mixed number below there is a matching letter. Convert all of the numbers into a form so that you can order the numbers from smallest to largest. If you order the numbers correctly, the matching letters will spell out the answer to the riddle.

2.123	T
$2\frac{1}{2}$	R
1.245	M
0.520	S
$\frac{6}{10}$	T
1.875	S
$\frac{9}{10}$	E
2.405	O
$1\frac{1}{4}$	O
$1\frac{1}{2}$	T
0.250	I
$2\frac{3}{4}$	E
$\frac{6}{8}$	H
0.340	T
$\frac{1}{2}$	A
2.645	I
$\frac{2}{5}$	H
$2\frac{4}{5}$	S
2.950	!
1.350	S

0.250	0.250	I
0.340	0.340	T
$\frac{2}{5}$	0.400	H
$\frac{1}{2}$	0.500	A
0.520	0.520	S
$\frac{6}{10}$	0.600	T
$\frac{6}{8}$	0.750	H
$\frac{9}{10}$	0.900	E
1.245	1.245	M
$1\frac{1}{4}$	1.250	O
1.350	1.350	S
$1\frac{1}{2}$	1.500	T
1.875	1.875	S
2.123	2.123	T
2.405	2.405	O
$2\frac{1}{2}$	2.500	R
2.645	2.645	I
$2\frac{3}{4}$	2.750	E
$2\frac{4}{5}$	2.800	S
2.950	2.950	!

Why is the library the tallest room in the school?

IT HAS THE MOST STORIES!

LESSON **4-5**

Practice A

Equivalent Fractions

Find two equivalent fractions for each fraction. Possible answers are given.

1. $\frac{1}{2}$ $\frac{2}{4}, \frac{3}{6}$
2. $\frac{1}{4}$ $\frac{2}{8}, \frac{3}{12}$
3. $\frac{1}{3}$ $\frac{2}{6}, \frac{3}{9}$
4. $\frac{1}{5}$ $\frac{2}{10}, \frac{3}{15}$
5. $\frac{2}{3}$ $\frac{4}{6}, \frac{6}{9}$
6. $\frac{3}{4}$ $\frac{6}{8}, \frac{9}{12}$
7. $\frac{1}{6}$ $\frac{2}{12}, \frac{3}{18}$
8. $\frac{2}{5}$ $\frac{4}{10}, \frac{6}{15}$
9. $\frac{10}{20}$ $\frac{5}{10}, \frac{1}{2}$

Find the missing numbers that make the fractions equivalent.

10. $\frac{2}{4} = \frac{?}{8}$ 4
11. $\frac{1}{3} = \frac{?}{9}$ 3
12. $\frac{10}{5} = \frac{?}{1}$ 2
13. $\frac{14}{4} = \frac{?}{2}$ 7
14. $\frac{5}{6} = \frac{?}{12}$ 10
15. $\frac{2}{7} = \frac{?}{21}$ 6

Write each fraction in simplest form.

16. $\frac{2}{4}$ $\frac{1}{2}$
17. $\frac{3}{9}$ $\frac{1}{3}$
18. $\frac{8}{10}$ $\frac{4}{5}$
19. $\frac{4}{6}$ $\frac{2}{3}$

20. A recipe calls for $\frac{1}{2}$ cup of sugar and $\frac{4}{8}$ cup of flour. Does the recipe use more sugar or flour? Explain.

It uses the same amount of sugar and flour, because $\frac{1}{2} = \frac{4}{8}$.

21. Frank and Antoine's teams played the same number of baseball games this season. Frank's team won $\frac{2}{3}$ of its games, and Antoine's team won $\frac{8}{12}$ of its games. Which team won the most games this season? Explain.

They won the same amount of games, because $\frac{2}{3} = \frac{8}{12}$.

LESSON **4-5**

Practice B

Equivalent Fractions

Find two equivalent fractions for each fraction. Possible answers are given.

1. $\frac{3}{6}$ $\frac{1}{2}, \frac{6}{12}$
2. $\frac{4}{7}$ $\frac{8}{14}, \frac{12}{21}$
3. $\frac{11}{13}$ $\frac{22}{26}, \frac{33}{39}$
4. $\frac{2}{15}$ $\frac{4}{30}, \frac{6}{45}$
5. $\frac{5}{14}$ $\frac{10}{28}, \frac{15}{42}$
6. $\frac{8}{9}$ $\frac{16}{18}, \frac{24}{27}$
7. $\frac{2}{21}$ $\frac{4}{42}, \frac{6}{63}$
8. $\frac{24}{48}$ $\frac{1}{2}, \frac{8}{16}$
9. $\frac{25}{100}$ $\frac{1}{4}, \frac{50}{200}$

Find the missing numbers that make the fractions equivalent.

10. $\frac{4}{7} = \frac{?}{28}$ 16
11. $\frac{2}{9} = \frac{?}{54}$ 12
12. $\frac{36}{4} = \frac{?}{1}$ 9
13. $\frac{56}{8} = \frac{?}{2}$ 14
14. $1\frac{3}{5} = \frac{?}{25}$ 40
15. $1\frac{4}{7} = \frac{?}{42}$ 66

Write each fraction in simplest form.

16. $\frac{15}{25}$ $\frac{3}{5}$
17. $\frac{8}{36}$ $\frac{2}{9}$
18. $\frac{12}{18}$ $\frac{2}{3}$
19. $\frac{10}{24}$ $\frac{5}{12}$

20. Billy had 24 trading cards. He gave 7 of his cards to Miko and 9 of his cards to Teri. What fraction of his original 24 cards does Billy have left? Write two equivalent fractions for that amount.

$\frac{8}{24}$; possible equivalent fractions: $\frac{4}{12}, \frac{2}{6}$

21. Beth and Kristine ride their bikes to school in the morning. Beth has to ride $1\frac{7}{32}$ miles. Kristine has to ride $\frac{39}{32}$ miles. Who rides the farthest to reach school? Explain.

They ride the same distance, because $1\frac{7}{32} = \frac{39}{32}$ miles.

LESSON **4-5** **Practice C**

Equivalent Fractions

Find two equivalent fractions for each fraction. **Possible answers are given.**

1. $\frac{10}{15}$ **$\frac{20}{30}$, $\frac{2}{3}$**
2. $\frac{7}{18}$ **$\frac{14}{36}$, $\frac{21}{54}$**
3. $\frac{5}{16}$ **$\frac{10}{32}$, $\frac{15}{48}$**
4. $\frac{9}{23}$ **$\frac{18}{46}$, $\frac{27}{69}$**
5. $\frac{14}{27}$ **$\frac{28}{54}$, $\frac{42}{81}$**
6. $\frac{155}{255}$ **$\frac{31}{51}$, $\frac{310}{510}$**
7. $\frac{62}{80}$ **$\frac{31}{40}$, $\frac{124}{160}$**
8. $\frac{21}{77}$ **$\frac{3}{11}$, $\frac{42}{154}$**
9. $\frac{94}{102}$ **$\frac{47}{51}$, $\frac{188}{204}$**

Find the missing numbers that make the fractions equivalent.

10. $\frac{2}{21} = \frac{?}{42}$ **4**
11. $\frac{28}{56} = \frac{?}{14}$ **7**
12. $\frac{15}{13} = \frac{?}{39}$ **45**
13. $1\frac{10}{12} = \frac{?}{6}$ **11**
14. $2\frac{3}{8} = \frac{?}{16}$ **38**
15. $2\frac{4}{6} = \frac{?}{3}$ **8**

Write each fraction in simplest form. Show two ways to simplify.

16. $\frac{18}{30}$ **$\frac{3}{5}$**
17. $\frac{16}{24}$ **$\frac{2}{3}$**
18. $\frac{16}{48}$ **$\frac{1}{3}$**
19. $\frac{12}{39}$ **$\frac{4}{13}$**

20. A theater manager has the same number of tickets available for every show. She sold $\frac{27}{45}$ of the tickets for Monday night's show. For Tuesday night's show, she sold $\frac{108}{180}$ of the tickets, and for Wednesday she sold $\frac{79}{135}$ of the tickets. For which two shows did she sell the same number of tickets? Explain.

Monday and Tuesday, because $\frac{27}{45} = \frac{108}{180}$

21. What do the fractions $\frac{3}{17}$, $\frac{5}{7}$, $\frac{11}{23}$, and $\frac{83}{97}$ have in common? Why is it impossible to write those fractions in simplest form?

All of the numbers in the fractions are prime; there are no numbers that divide both the numerator and the denominator in each fraction.

LESSON **4-5** **Reteach**

Equivalent Fractions

Fractions that have the same value are equivalent fractions.

$\frac{3}{4} = \frac{6}{8}$

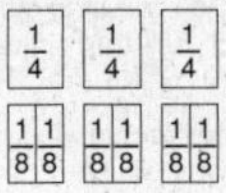

$\frac{3}{4}$ and $\frac{6}{8}$ are equivalent fractions.

You can use fraction strips to help you find equivalent fractions.

To solve $\frac{2}{3} = \frac{?}{12}$, first use fraction strips to model the first fraction.

Then use $\frac{1}{12}$ fraction pieces to make a length as long as the $\frac{2}{3}$ strip.

You need eight $\frac{1}{12}$ pieces, so $\frac{2}{3} = \frac{8}{12}$.

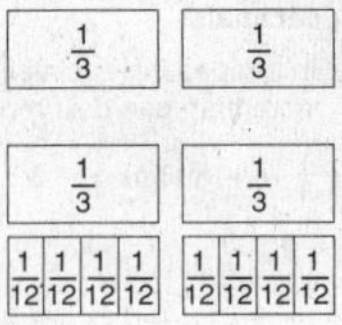

Find the missing number that makes the fractions equivalent.

1. $\frac{3}{4} = \frac{?}{12}$ **9**
2. $\frac{8}{10} = \frac{?}{5}$ **4**
3. $\frac{6}{9} = \frac{?}{3}$ **2**
4. $\frac{5}{6} = \frac{?}{12}$ **10**

A fraction is in simplest form when the GCF of the numerator and the denominator is 1.

To write $\frac{4}{6}$ in simplest form, divide the numerator and the denominator by their GCF, 2.

$4 \div 2 = 2$

$6 \div 2 = 3$

So, $\frac{4}{6}$ in simplest form is $\frac{2}{3}$.

Write each fraction in simplest form.

5. $\frac{3}{9}$ **$\frac{1}{3}$**
6. $\frac{12}{16}$ **$\frac{3}{4}$**
7. $\frac{14}{18}$ **$\frac{7}{9}$**
8. $\frac{8}{20}$ **$\frac{2}{5}$**

LESSON **4-5** **Challenge**

Match That Fraction!

Write the fraction that the shaded part of each figure represents. Then match equivalent fractions by drawing lines between numbered and lettered figures.

1. 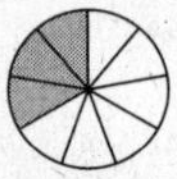

Fraction: **$\frac{3}{9}$**

2. 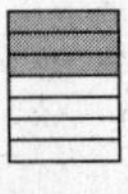

Fraction: **$\frac{3}{7}$**

3. 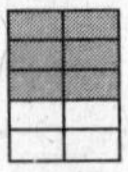

Fraction: **$\frac{6}{10}$**

4. 

Fraction: **$\frac{4}{16}$**

5. 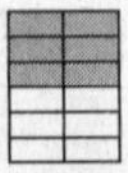

Fraction: **$\frac{6}{12}$**

A. 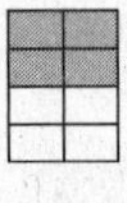

Fraction: **$\frac{4}{8}$**

B. 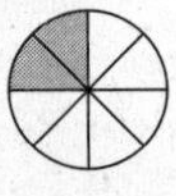

Fraction: **$\frac{2}{8}$**

C. 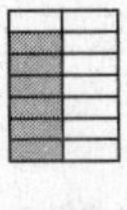

Fraction: **$\frac{6}{14}$**

D. 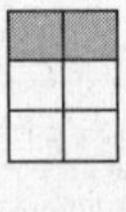

Fraction: **$\frac{2}{6}$**

E. 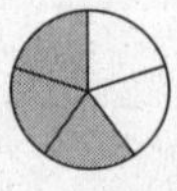

Fraction: **$\frac{3}{5}$**

LESSON **4-5** **Problem Solving**

Equivalent Fractions

About 60 million Americans exercise 100 times or more each year. Their top activities and the fraction of those 60 million people who did them are shown on the circle graph. Use the graph to answer the questions.

Exercise in the U.S.

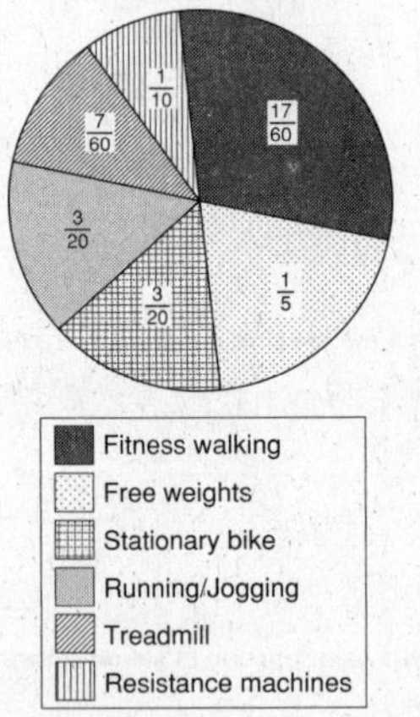

1. Which two activities did the same number of people use to keep in shape?

stationary bike and running/jogging

2. Which activity had the most participants? Write an equivalent fraction for that activity's participants.

fitness walking; possible answer: $\frac{34}{120}$

3. Which activity had the fewest participants? Write two equivalent fractions for that activity's participants.

resistance machines; possible answers: $\frac{2}{20}$, $\frac{3}{30}$

Circle the letter of the correct answer.

4. Which activity did $\frac{3}{15}$ of the people use to exercise?
   - (A) free weights (circled)
   - B treadmill
   - C fitness walking
   - D stationary bike

5. Which activity did $\frac{35}{300}$ of the people use to stay healthy?
   - F running/jogging
   - G resistance machines
   - H free weights
   - (J) treadmill (circled)

6. An average-sized person can burn about $6\frac{1}{2}$ calories a minute while riding a bike. Which of the following is equivalent to that amount?
   - A $1\frac{2}{2}$
   - (C) $6\frac{2}{4}$ (circled)
   - B $5\frac{6}{2}$
   - D $6\frac{2}{6}$

7. An average-sized person can burn about 11.25 calories a minute while jogging. Which of the following is not equivalent to that amount?
   - F $11\frac{1}{4}$
   - H $11\frac{2}{8}$
   - (G) $11\frac{1}{2}$ (circled)
   - J $11\frac{3}{12}$

LESSON 4-5
Reading Strategies
Use a Visual Model

Each fraction has the same amount shaded. Fractions of equal value are called **equivalent fractions.**

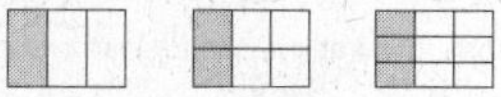

$\frac{1}{3} = \frac{2}{6} = \frac{3}{9}$

Answer the following questions.

1. What fractional part of rectangle A is shaded? $\frac{1}{3}$
2. What fractional part of rectangle B is shaded? $\frac{2}{6}$
3. What fractional part of rectangle C is shaded? $\frac{3}{9}$
4. Write two equivalent fractions for $\frac{1}{3}$. $\frac{2}{6}$ and $\frac{3}{9}$

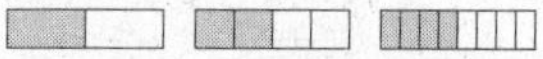

5. What fractional part of rectangle D is shaded? $\frac{1}{2}$
6. Shade an equivalent fractional part of rectangle E.
7. Shade an equivalent fractional part of rectangle F.
8. Write two equivalent fractions for $\frac{1}{2}$. $\frac{2}{4}$ and $\frac{4}{8}$

LESSON 4-5
Puzzles, Twisters & Teasers
The Best Day in School

Your goal is to find all of the fractions in the grid which are equivalent.

Black out any pair or set of three fractions that are equivalent. Read the letters in the remaining boxes in order from left to right, top to bottom, to solve the riddle.

~~$\frac{6}{10}$ W~~	~~$\frac{12}{32}$ C~~	$\frac{45}{55}$ D	~~$\frac{7}{14}$ A~~	~~$\frac{18}{45}$ O~~	~~$\frac{21}{27}$ L~~
~~$\frac{18}{48}$ E~~	$\frac{2}{14}$ E	~~$\frac{8}{16}$ V~~	~~$\frac{6}{21}$ X~~	~~$\frac{36}{40}$ M~~	~~$\frac{12}{26}$ F~~
~~$\frac{20}{25}$ G~~	~~$\frac{15}{18}$ T~~	$\frac{9}{27}$ L	~~$\frac{5}{40}$ S~~	$\frac{2}{18}$ I	~~$\frac{15}{25}$ T~~
~~$\frac{21}{36}$ T~~	~~$\frac{16}{20}$ H~~	~~$\frac{46}{92}$ S~~	~~$\frac{35}{45}$ P~~	~~$\frac{18}{39}$ N~~	~~$\frac{21}{27}$ K~~
~~$\frac{14}{24}$ Y~~	$\frac{4}{44}$ G	$\frac{8}{18}$ H	~~$\frac{6}{24}$ E~~	~~$\frac{8}{28}$ B~~	$\frac{24}{36}$ T
$\frac{14}{20}$ E	~~$\frac{6}{24}$ O~~	~~$\frac{4}{10}$ U~~	$\frac{24}{28}$ D	~~$\frac{55}{66}$ I~~	~~$\frac{90}{100}$ E~~

What happened when the power went off at school?

The students were DELIGHTED!

LESSON 4-6
Practice A
Mixed Numbers and Improper Fractions

Determine whether each fraction is proper or improper.

1. $\frac{1}{4}$ proper
2. $\frac{3}{4}$ proper
3. $\frac{7}{4}$ improper
4. $\frac{9}{3}$ improper
5. $\frac{2}{3}$ proper
6. $\frac{6}{2}$ improper

Write each mixed number as an improper fraction.

7. $1\frac{1}{2}$ $\frac{3}{2}$
8. $1\frac{1}{3}$ $\frac{4}{3}$
9. $2\frac{1}{2}$ $\frac{5}{2}$
10. $1\frac{1}{4}$ $\frac{5}{4}$
11. $1\frac{2}{3}$ $\frac{5}{3}$
12. $2\frac{1}{3}$ $\frac{7}{3}$

Write each improper fraction as a mixed number or whole number. Tell whether your answer is a mixed number or whole number.

13. $\frac{4}{1}$ 4; whole number
14. $\frac{6}{3}$ 2; whole number
15. $\frac{4}{3}$ $1\frac{1}{3}$; mixed number
16. $\frac{5}{2}$ $2\frac{1}{2}$; mixed number
17. $\frac{7}{5}$ $1\frac{2}{5}$; mixed number
18. $\frac{3}{2}$ $1\frac{1}{2}$; mixed number

19. Mickey timed her speech at $\frac{12}{5}$ minutes. She cannot go over 2 minutes. Should she add more to her speech, or cut some lines? Explain.
She should cut some lines because $\frac{12}{5} = 2\frac{2}{5}$ minutes, which is longer than 2 minutes.

20. In a frog-jumping contest, Jim's frog hopped $2\frac{1}{3}$ feet. Matthew's frog hopped $1\frac{2}{3}$ feet. Whose frog hopped $\frac{7}{3}$ feet?
Jim's frog

LESSON 4-6
Practice B
Mixed Numbers and Improper Fractions

Write each mixed number as an improper fraction.

1. $3\frac{1}{2}$ $\frac{7}{2}$
2. $2\frac{1}{3}$ $\frac{7}{3}$
3. $5\frac{1}{4}$ $\frac{21}{4}$
4. $1\frac{3}{7}$ $\frac{10}{7}$
5. $3\frac{3}{4}$ $\frac{15}{4}$
6. $4\frac{1}{3}$ $\frac{13}{3}$
7. $2\frac{3}{5}$ $\frac{13}{5}$
8. $3\frac{5}{6}$ $\frac{23}{6}$
9. $7\frac{1}{3}$ $\frac{22}{3}$

Write each improper fraction as a mixed number or whole number. Tell whether your answer is a mixed number or whole number.

10. $\frac{17}{3}$ $5\frac{2}{3}$; mixed number
11. $\frac{40}{8}$ 5; whole number
12. $\frac{48}{7}$ $6\frac{6}{7}$; mixed number
13. $\frac{33}{10}$ $3\frac{3}{10}$; mixed number
14. $\frac{50}{8}$ $6\frac{1}{4}$; mixed number
15. $\frac{83}{9}$ $9\frac{2}{9}$; mixed number
16. $\frac{104}{8}$ 13; whole number
17. $\frac{121}{6}$ $20\frac{1}{6}$; mixed number
18. $\frac{78}{11}$ $7\frac{1}{11}$; mixed number

19. The hotel ordered an extra-long rug for a hallway that is $\frac{123}{2}$ feet long. What is the rug's length in feet and inches? Remember, 1 foot = 12 inches.
61 feet and 6 inches

20. During this year's football-throwing contest, John threw the ball $49\frac{2}{3}$ feet. Sharon threw the ball 51 feet. Who threw the ball $\frac{153}{3}$ feet?
Sharon

LESSON 4-6
Practice C
Mixed Numbers and Improper Fractions

Write each mixed number as an improper fraction.

1. $2\frac{2}{5}$ $\frac{12}{5}$
2. $1\frac{5}{10}$ $\frac{15}{10}$
3. $6\frac{3}{4}$ $\frac{27}{4}$
4. $2\frac{5}{9}$ $\frac{23}{9}$
5. $14\frac{2}{3}$ $\frac{44}{3}$
6. $2\frac{9}{11}$ $\frac{31}{11}$

Write each improper fraction as a mixed number or whole number. Tell whether your answer is a mixed number or whole number.

7. $\frac{161}{7}$ 23; whole number
8. $\frac{73}{4}$ $18\frac{1}{4}$; mixed number
9. $\frac{52}{9}$ $5\frac{7}{9}$; mixed number
10. $\frac{225}{9}$ 25; whole number
11. $\frac{234}{11}$ $21\frac{3}{11}$; mixed number
12. $\frac{127}{12}$ $10\frac{7}{12}$; mixed number

Replace each shape with a number that will make the equation correct.

13. $▲\frac{2}{7} = \frac{93}{♦}$ 13; 7
14. $▲\frac{3}{8} = \frac{67}{♦}$ 8; 8
15. $▲\frac{4}{5} = \frac{64}{♦}$ 12; 5
16. $43\frac{▲}{15} = \frac{647}{♦}$ 2; 15
17. $13\frac{▲}{9} = \frac{124}{♦}$ 7; 9
18. $14\frac{▲}{8} = \frac{113}{♦}$ 1; 8

19. Amy made a video that is 5,400 seconds long. What is the video's running time in hours and minutes?

 1 hour and 30 minutes

20. Justin accidentally erased the numerator of the fraction part of a mixed number. He knows that its whole number part is 45 and the denominator of the fraction part is 9. He also knows that when the mixed number is written as an improper fraction, its numerator is 412. What is the numerator that Justin erased?

 7

LESSON 4-6
Reteach
Mixed Numbers and Improper Fractions

A proper fraction is a fraction whose numerator is less than its denominator.

$\frac{2}{3}$, $\frac{1}{4}$, and $\frac{2}{7}$ are examples of proper fractions.

An improper fraction is a fraction whose numerator is greater than or equal to its denominator.

$\frac{3}{2}$, $\frac{8}{3}$, and $\frac{5}{5}$ are examples of improper fractions.

Some improper fractions can be written as mixed numbers.

To write $\frac{7}{4}$ as a mixed number, draw circles divided into $\frac{1}{4}$ sections.

Then shade in 7 of the $\frac{1}{4}$ sections.

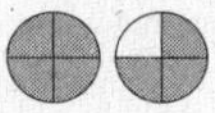

There is one circle and $\frac{3}{4}$ of a circle shaded.

So, $\frac{7}{4} = 1\frac{3}{4}$.

Write each improper fraction as a mixed number.

1. $\frac{14}{3}$ $4\frac{2}{3}$
2. $\frac{11}{2}$ $5\frac{1}{2}$
3. $\frac{15}{4}$ $3\frac{3}{4}$
4. $\frac{19}{6}$ $3\frac{1}{6}$

Mixed numbers can be written as improper fractions.

To write $2\frac{1}{3}$ as an improper fraction, draw 3 circles. Divide each circle into $\frac{1}{3}$ sections. Next, shade in 2 whole circles and one $\frac{1}{3}$ section of the last circle.

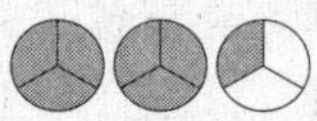

Then find the total number of $\frac{1}{3}$ sections that are shaded.

Seven $\frac{1}{3}$ sections are shaded, so $2\frac{1}{3} = \frac{7}{3}$.

Write each mixed number as an improper fraction.

5. $3\frac{1}{4}$ $\frac{13}{4}$
6. $5\frac{2}{3}$ $\frac{17}{3}$
7. $4\frac{1}{2}$ $\frac{9}{2}$
8. $1\frac{5}{6}$ $\frac{11}{6}$

LESSON 4-6
Challenge
Mixed Up Mixed Numbers

Find the whole number and fraction in the box to make a mixed number to match each improper fraction listed below. Cross out each whole number and fraction as you use it. When you have finished, the unused items in the box will form the mixed number that answers the question at the bottom of the page.

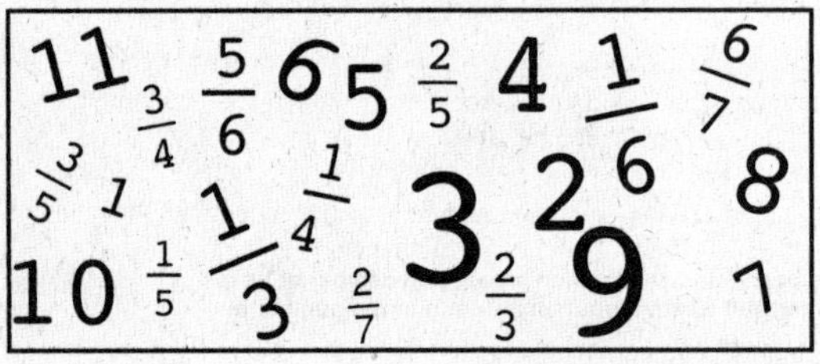

1. $\frac{7}{3} =$ $2\frac{1}{3}$
2. $\frac{20}{3} =$ $6\frac{2}{3}$
3. $\frac{21}{4} =$ $5\frac{1}{4}$
4. $\frac{31}{4} =$ $7\frac{3}{4}$
5. $\frac{41}{5} =$ $8\frac{1}{5}$
6. $\frac{57}{5} =$ $11\frac{2}{5}$
7. $\frac{59}{6} =$ $9\frac{5}{6}$
8. $\frac{19}{6} =$ $3\frac{1}{6}$
9. $\frac{13}{7} =$ $1\frac{6}{7}$
10. $\frac{72}{7} =$ $10\frac{2}{7}$

How old is the moon? $4\frac{3}{5}$ billion years old!

LESSON 4-6
Problem Solving
Mixed Numbers and Improper Fractions

Write the correct answer.

1. If stretched end-to-end, the total length of the blood vessels inside your body could wrap around Earth's equator $\frac{5}{2}$ times! Write this fact as a mixed number.

 $2\frac{1}{2}$ times

2. In 2000, the average 12-year-old child in the United States earned an allowance of 9 dollars and $\frac{7}{25}$ cents a week. Write this amount as an improper fraction and a decimal.

 $\$\frac{232}{25}$; \$9.28

3. The normal body temperature for a rattlesnake is between $53\frac{3}{5}$°F and $64\frac{2}{5}$°F. Write this range as improper fractions.

 $\frac{268}{5}$°F to $\frac{322}{5}$°F

4. A professional baseball can weigh no less than $\frac{45}{9}$ ounces and no more than $\frac{21}{4}$ ounces. Write this range as mixed numbers.

 5 ounces to $5\frac{1}{4}$ ounces

Circle the letter of the correct answer.

5. Betty needs a piece of lumber that is $\frac{14}{3}$ feet long. Which size should she look for at the hardware store?

 A $3\frac{1}{3}$ feet
 B $3\frac{1}{4}$ feet
 (C) $4\frac{2}{3}$ feet
 D $4\frac{1}{4}$ feet

6. What operations are used to change a mixed number to an improper fraction?

 (F) multiplication and addition
 G division and subtraction
 H division and addition
 J multiplication and subtraction

7. Adult bees only eat nectar, the substance in flowers used to make honey. A bee could fly 4 million miles on the energy it would get from eating $\frac{9}{2}$ liters of nectar. What is this amount of nectar written as a mixed number.

 A $9\frac{1}{2}$ liters
 (B) $4\frac{1}{2}$ liters
 C $4\frac{1}{9}$ liters
 D $2\frac{1}{2}$ liters

8. An astronaut who weighs 250 pounds on Earth would weigh $41\frac{1}{2}$ pounds on the moon. What is the astronaut's moon weight written as an improper fraction?

 F $\frac{41}{2}$ pounds
 G $\frac{42}{2}$ pounds
 H $\frac{82}{2}$ pounds
 (J) $\frac{83}{2}$ pounds

LESSON **4-6**
Reading Strategies
Focus on Vocabulary

A **mixed number** is made up of a whole number and a fraction.

Read: ⟶ two and one-half

Write: ⟶ $2\frac{1}{2}$

In an **improper fraction,** the numerator is greater than or equal to the denominator.

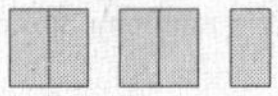

Read: ⟶ five halves

Write: ⟶ $\frac{5}{2}$

Answer the following questions.

1. What is a mixed number?

   a number made up of a whole number and a fraction

2. Why do you think it is called a "mixed" number?

   Sample answer: because a whole number and a fraction are two different types of numbers

3. What is an improper fraction?

   a fraction that has a numerator greater than the denominator

4. Why do you think it is called an "improper" fraction?

   Sample answer: because it is actually more than a fraction.

Write "mixed number" or "improper fraction" for each number.

5. $1\frac{4}{5}$ mixed number
6. $\frac{15}{8}$ improper fraction
7. $\frac{20}{17}$ improper fraction
8. $3\frac{2}{3}$ mixed number

LESSON **4-6**
Puzzles, Twisters & Teasers
Surfing the Web

When you send information over the Internet it does not go in a straight path. Get your information from your computer to your friend by finding out the path it will take from Send to Receive.

You must always move from a smaller number to a larger number. You may move left, right, up and down only (no diagonal moves).

Send

$\frac{1}{8}$	$\frac{1}{16}$	$\frac{7}{3}$	$2\frac{7}{8}$	$\frac{17}{5}$	$2\frac{15}{16}$
$\frac{3}{16}$	$1\frac{1}{2}$	$1\frac{3}{5}$	$1\frac{1}{8}$	$3\frac{3}{4}$	$\frac{25}{8}$
$\frac{1}{8}$	$\frac{5}{4}$	$1\frac{1}{2}$	$\frac{16}{4}$	$\frac{33}{8}$	$4\frac{1}{16}$
$\frac{25}{5}$	$\frac{47}{9}$	$4\frac{5}{6}$	$\frac{9}{2}$	$4\frac{3}{16}$	$\frac{31}{8}$
$5\frac{1}{6}$	$5\frac{1}{3}$	$\frac{9}{2}$	$4\frac{1}{3}$	$4\frac{1}{8}$	$4\frac{8}{9}$
$\frac{11}{2}$	$\frac{35}{6}$	$\frac{25}{4}$	$6\frac{2}{7}$	$6\frac{1}{4}$	$\frac{27}{5}$
$\frac{20}{3}$	$5\frac{5}{9}$	$\frac{18}{3}$	$\frac{14}{2}$	$6\frac{1}{2}$	$6\frac{7}{11}$
$\frac{31}{4}$	$\frac{58}{7}$	$7\frac{6}{7}$	$8\frac{5}{8}$	$8\frac{7}{16}$	$\frac{23}{3}$
$8\frac{1}{8}$	$\frac{17}{2}$	$\frac{88}{11}$	$\frac{37}{4}$	$\frac{9}{1}$	$\frac{70}{7}$
$9\frac{1}{16}$	$9\frac{1}{8}$	$9\frac{3}{16}$	$9\frac{2}{5}$	$\frac{59}{6}$	$10\frac{1}{5}$

Receive

LESSON **4-7**
Practice A
Comparing and Ordering Fractions

Compare. Write <, >, or =.

1. $\frac{1}{6} < \frac{2}{6}$
2. $\frac{1}{3} < \frac{2}{3}$
3. $\frac{2}{4} = \frac{1}{2}$
4. $\frac{1}{2} < \frac{3}{4}$
5. $\frac{4}{6} = \frac{2}{3}$
6. $\frac{1}{5} < \frac{2}{3}$
7. $\frac{1}{2} > \frac{2}{5}$
8. $\frac{3}{4} > \frac{3}{5}$
9. $\frac{1}{9} < \frac{1}{8}$

Order the fractions from least to greatest.

10. $\frac{7}{9}, \frac{1}{9}, \frac{2}{9}$ — $\frac{1}{9}, \frac{2}{9}, \frac{7}{9}$
11. $\frac{1}{3}, \frac{1}{4}, \frac{2}{3}$ — $\frac{1}{4}, \frac{1}{3}, \frac{2}{3}$
12. $\frac{2}{5}, \frac{1}{2}, \frac{1}{10}$ — $\frac{1}{10}, \frac{2}{5}, \frac{1}{2}$

Order the fractions from greatest to least.

13. $\frac{1}{5}, \frac{3}{5}, \frac{2}{5}$ — $\frac{3}{5}, \frac{2}{5}, \frac{1}{5}$
14. $\frac{2}{7}, \frac{6}{7}, \frac{3}{6}$ — $\frac{6}{7}, \frac{3}{6}, \frac{2}{7}$
15. $\frac{1}{6}, \frac{2}{3}, \frac{5}{6}$ — $\frac{5}{6}, \frac{2}{3}, \frac{1}{6}$

16. The Mountain Cats and the Pirates have both played the same number of games. The Mountain Cats have won $\frac{1}{5}$ of its games, and the Pirates have won $\frac{1}{4}$ of its games. Which team has the most wins?

    the Pirates

17. Each week, Kelly saves $\frac{1}{2}$ of her allowance. She spends $\frac{1}{3}$ of her allowance on food, and $\frac{1}{6}$ of it on games and movies. What does Kelly do with the largest part of her allowance?

    She saves it.

LESSON **4-7**
Practice B
Comparing and Ordering Fractions

Compare. Write <, >, or =.

1. $\frac{4}{7} < \frac{3}{5}$
2. $\frac{1}{8} < \frac{2}{3}$
3. $\frac{1}{4} < \frac{2}{5}$
4. $\frac{7}{8} > \frac{5}{6}$
5. $\frac{18}{24} = \frac{3}{4}$
6. $\frac{4}{5} > \frac{8}{12}$

Order the fractions from least to greatest.

7. $\frac{1}{2}, \frac{2}{5}, \frac{1}{3}$ — $\frac{1}{3}, \frac{2}{5}, \frac{1}{2}$
8. $\frac{2}{5}, \frac{3}{4}, \frac{2}{3}$ — $\frac{2}{5}, \frac{2}{3}, \frac{3}{4}$
9. $\frac{3}{7}, \frac{5}{6}, \frac{4}{5}$ — $\frac{3}{7}, \frac{4}{5}, \frac{5}{6}$
10. $\frac{5}{9}, \frac{3}{7}, \frac{2}{3}$ — $\frac{3}{7}, \frac{5}{9}, \frac{2}{3}$
11. $\frac{3}{8}, \frac{2}{7}, \frac{3}{5}$ — $\frac{2}{7}, \frac{3}{8}, \frac{3}{5}$
12. $\frac{2}{7}, \frac{1}{8}, \frac{2}{5}$ — $\frac{1}{8}, \frac{2}{7}, \frac{2}{5}$

Order the fractions from greatest to least.

13. $\frac{1}{6}, \frac{2}{7}, \frac{1}{5}$ — $\frac{2}{7}, \frac{1}{5}, \frac{1}{6}$
14. $\frac{3}{7}, \frac{4}{9}, \frac{2}{3}$ — $\frac{2}{3}, \frac{4}{9}, \frac{3}{7}$
15. $\frac{2}{5}, \frac{3}{10}, \frac{2}{3}$ — $\frac{2}{3}, \frac{2}{5}, \frac{3}{10}$
16. $\frac{4}{5}, \frac{7}{10}, \frac{1}{12}$ — $\frac{4}{5}, \frac{7}{10}, \frac{1}{12}$
17. $\frac{3}{8}, \frac{3}{4}, \frac{4}{9}$ — $\frac{3}{4}, \frac{4}{9}, \frac{3}{8}$
18. $\frac{4}{7}, \frac{3}{5}, \frac{5}{6}$ — $\frac{5}{6}, \frac{3}{5}, \frac{4}{7}$

19. David ran $4\frac{1}{4}$ miles, Shane ran $4\frac{1}{2}$ miles, and Matt ran $4\frac{5}{8}$ miles. Who ran the farthest?

    Matt

20. Darius and Anita both took the same test. Darius answered $\frac{5}{6}$ of the questions correctly, and Anita answered $\frac{6}{7}$ correctly. Who got the higher score on the test?

    Anita

LESSON 4-7
Practice C
Comparing and Ordering Fractions

Compare. Write <, >, or =.

1. $\frac{4}{9}$ [<] $\frac{5}{7}$
2. $\frac{5}{12}$ [<] $\frac{13}{22}$
3. $\frac{13}{15}$ [>] $\frac{6}{7}$
4. $\frac{16}{35}$ [<] $\frac{12}{20}$
5. $\frac{13}{16}$ [=] $\frac{39}{48}$
6. $\frac{9}{11}$ [<] $\frac{7}{8}$

Order the fractions from least to greatest.

7. $\frac{3}{8}, \frac{2}{7}, \frac{4}{9}$ **$\frac{2}{7}, \frac{3}{8}, \frac{4}{9}$**
8. $\frac{6}{11}, \frac{4}{5}, \frac{3}{4}$ **$\frac{6}{11}, \frac{3}{4}, \frac{4}{5}$**
9. $\frac{2}{7}, \frac{3}{9}, \frac{6}{7}$ **$\frac{2}{7}, \frac{3}{9}, \frac{6}{7}$**
10. $\frac{9}{10}, \frac{5}{6}, \frac{2}{3}$ **$\frac{2}{3}, \frac{5}{6}, \frac{9}{10}$**
11. $\frac{13}{15}, \frac{8}{9}, \frac{5}{8}$ **$\frac{5}{8}, \frac{13}{15}, \frac{8}{9}$**
12. $\frac{6}{7}, \frac{7}{10}, \frac{12}{13}$ **$\frac{7}{10}, \frac{6}{7}, \frac{12}{13}$**

Order the fractions from greatest to least.

13. $\frac{2}{3}, \frac{5}{6}, \frac{4}{7}$ **$\frac{5}{6}, \frac{2}{3}, \frac{4}{7}$**
14. $\frac{7}{10}, \frac{5}{7}, \frac{11}{13}$ **$\frac{11}{13}, \frac{5}{7}, \frac{7}{10}$**
15. $\frac{9}{15}, \frac{5}{8}, \frac{10}{13}$ **$\frac{10}{13}, \frac{5}{8}, \frac{9}{15}$**
16. $\frac{11}{17}, \frac{5}{8}, \frac{7}{9}$ **$\frac{7}{9}, \frac{11}{17}, \frac{5}{8}$**
17. $\frac{5}{9}, \frac{6}{7}, \frac{5}{6}$ **$\frac{6}{7}, \frac{5}{6}, \frac{5}{9}$**
18. $\frac{2}{9}, \frac{3}{8}, \frac{4}{7}$ **$\frac{4}{7}, \frac{3}{8}, \frac{2}{9}$**

19. A group of students collected $1,724 for the Red Cross. Bob collected $\frac{7}{24}$ of the money, Grady collected $\frac{3}{8}$, and Vania collected $\frac{1}{3}$. Who collected the most money?

Grady

20. The Gators soccer team plays the same number of games each year. The team won $\frac{9}{11}$ of its games in 2000, $\frac{81}{99}$ of its games in 2001, and $\frac{63}{77}$ of its games in 2002. In which year did the team win the most games?

The team won the same number of games each year.

LESSON 4-7
Reteach
Comparing and Ordering Fractions

$\frac{3}{4}$ and $\frac{2}{3}$ are unlike fractions because they have different denominators. To compare these fractions, graph the fractions on the same number line.

The fraction that is farther to the right on the number line is greater in value.

$\frac{9}{12} > \frac{8}{12}$, so $\frac{3}{4} > \frac{2}{3}$.

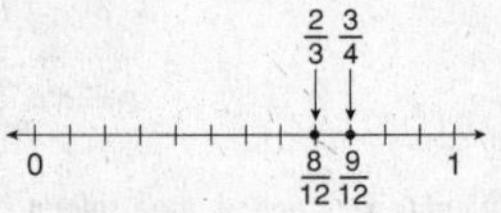

Use the number line to compare the fractions. Write < or >.

1. $\frac{1}{3}$ [<] $\frac{5}{6}$
2. $\frac{5}{10}$ [>] $\frac{1}{4}$
3. $\frac{11}{12}$ [>] $\frac{4}{6}$
4. $\frac{6}{9}$ [>] $\frac{1}{6}$

You can use number lines to order fractions from least to greatest.

To order $\frac{1}{2}, \frac{5}{6}$, and $\frac{3}{4}$, graph the values on a number line.

Then list the fractions as they appear from left to right. From least to greatest the fractions are $\frac{1}{2}, \frac{3}{4}$, and $\frac{5}{6}$.

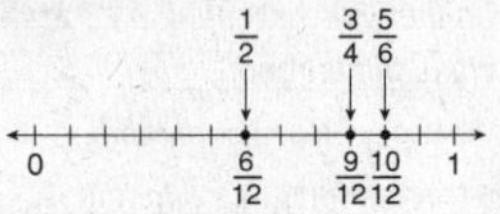

Use the number line to order the fractions from least to greatest.

5. $\frac{7}{12}, \frac{1}{4}, \frac{4}{8}$ **$\frac{1}{4}, \frac{4}{8}, \frac{7}{12}$**
6. $\frac{2}{3}, \frac{7}{12}, \frac{2}{8}$ **$\frac{2}{8}, \frac{7}{12}, \frac{2}{3}$**
7. $\frac{1}{2}, \frac{3}{4}, \frac{3}{9}$ **$\frac{3}{9}, \frac{1}{2}, \frac{3}{4}$**
8. $\frac{5}{12}, \frac{1}{3}, \frac{5}{6}$ **$\frac{1}{3}, \frac{5}{12}, \frac{5}{6}$**

LESSON 4-7
Challenge
Light as a Feather

The pygmy shrew is the lightest mammal on Earth. It weighs only 0.05 ounces. That's lighter than most feathers! Use the clues below to complete the table with each tiny animal's weight.

Lightest Mammals

Mammal	Weight (oz)
Pygmy shrew	$\frac{1}{20}$
Pipistrelle bat	$\frac{11}{100}$
Kitti's hog-nosed bat	$\frac{7}{100}$
Harvest mouse	$\frac{9}{50}$
Masked shrew	$\frac{2}{25}$

The smallest pygmy shrews are only $1\frac{1}{2}$ inches long!

1. Thousands of masked shrews live all over North America, but they are rarely seen. Although tiny, they have enormous appetites—often eating more than their body weight each day in bugs, slugs, and other crawly things. A masked shrew weighs more than a kitti's hog-nosed bat, but less than a pipistrelle bat.

2. Several different speices of harvest mice live in the United States. They are excellent climbers and spend much of the night scampering around in search of seeds and bugs to eat. During the day, they stay at home to avoid their worst enemies: hawks, owls, and snakes. A harvest mouse weighs more than any other animal listed in the table.

3. Kitti's hog-nosed bats live only in Asia. They spend most of the day hanging in deep, dark caves. At night, they leave the cave and fly around catching bugs to eat. About the size of a bumblebee, a kitti's hog-nosed bat weighs more than a pygmy shrew, but less than a masked shrew.

4. Pipistrelle bats are the smallest bats in the United States. They are unusual for bats, because they often fly in daylight to catch beetles and other bugs to eat. In winter, they hibernate in caves or other dark places. During that time, the bats are often covered with droplets of water, which sparkle and give them a pearly glow. A pipistrelle bat weighs less than a harvest mouse, but more than a masked shrew.

LESSON 4-7
Problem Solving
Comparing and Ordering Fractions

The table shows what fraction of Earth's total land area each of the continents makes up. Use the table to answer the questions.

Earth's Land

Continent	Fraction of Earth's Land
Africa	$\frac{1}{5}$
Antarctica	$\frac{1}{10}$
Asia	$\frac{3}{10}$
Australia	$\frac{1}{20}$
Europe	$\frac{7}{100}$
North America	$\frac{4}{25}$
South America	$\frac{6}{50}$

1. Which continent makes up most of Earth's land?

Asia

2. Which continent makes up the least part of Earth's land?

Australia

3. Explain how you would compare the part of Earth's total land area that Australia and Europe make up.

Change $\frac{1}{20}$ to $\frac{5}{100}$ and then compare it to $\frac{7}{100}$.

Circle the letter of the correct answer.

4. Which of these continents covers the greatest part of Earth's total land area?
   (A) North America
   B South America
   C Europe
   D Australia

5. Which of these continents covers the least part of Earth's total land area?
   F Africa
   G Antarctica
   H Asia
   (J) Australia

6. Which of the following lists shows the continents written in order from the greatest part of Earth's total land they cover to the least part?
   (A) Asia, Africa, North America
   B Africa, Asia, North America
   C Asia, South America, North America
   D North America, Asia, South America

7. Which of the following lists shows the continents written in order from the least part of Earth's total land they cover to the greatest part?
   F Antarctica, Europe, South America
   G South America, Antarctica, Europe
   (H) Australia, Europe, Antarctica
   J Antarctica, Europe, Australia

LESSON 4-7

Reading Strategies

Use a Graphic Aid

A number line can help you compare fractions.

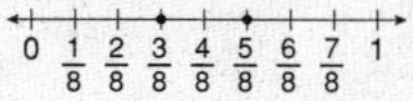

Use the number line and write >, <, or = for each.

1. $\frac{1}{8}$ [<] $\frac{5}{8}$
2. $\frac{5}{8}$ [>] $\frac{2}{8}$
3. $\frac{6}{8}$ [>] $\frac{4}{8}$

4. When you compare two fractions with the same denominator, how you can tell which one has the greater value?

Sample answer: The fraction with the greater numerator has the greater value.

A number line can help you compare fractions with different denominators.

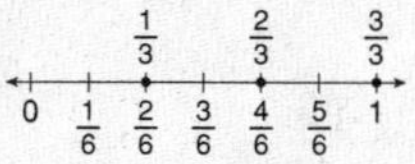

Use the number line and write >, <, or =.

5. $\frac{2}{6}$ [<] $\frac{2}{3}$
6. $\frac{5}{6}$ [>] $\frac{1}{3}$
7. $\frac{2}{3}$ [=] $\frac{4}{6}$
8. $\frac{3}{6}$ [<] $\frac{3}{3}$
9. $\frac{5}{6}$ [>] $\frac{2}{3}$
10. $\frac{1}{3}$ [>] $\frac{1}{6}$

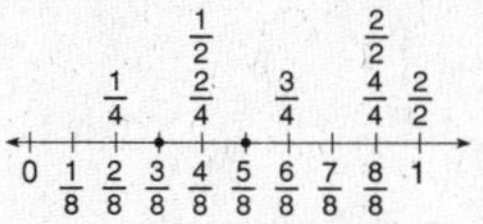

Use the number line and write >, <, or =.

11. $\frac{3}{8}$ [<] $\frac{3}{4}$
12. $\frac{3}{4}$ [<] $\frac{7}{8}$
13. $\frac{1}{2}$ [<] $\frac{5}{8}$
14. $\frac{2}{8}$ [=] $\frac{1}{4}$
15. $\frac{2}{4}$ [=] $\frac{4}{8}$
16. $\frac{6}{8}$ [>] $\frac{1}{2}$

LESSON 4-7

Puzzles, Twisters & Teasers

Celestial Fractions

Your goal is to pull fractions out of the "sky." Find the single largest fraction and the single smallest fraction in the cloud of fractions below.

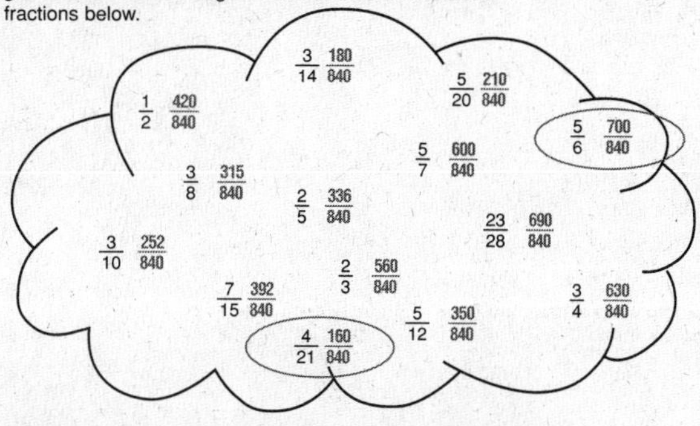

To solve the riddle, take the denominator of the largest fraction as given and find it in the rectangle below. Record the letter directly below that number in the first answer blank. Move clockwise, skip the next letter, and write the letter that comes after that in the next answer blank. Continue skipping and listing letters until you have the eight letters of the first word.

Find the denominator of the smallest fraction and repeat the process to find the second word.

How can you tell that your friend plays cards?

He **S H U F F L E S** when he **W A L K S**!

20 E	14 S	6 S	15 K	8 H	7 O
2 K					28 U
12 L	5 L	7 F	3 A	10 F	21 W

LESSON 4-8

Practice A

Adding and Subtracting with Like Denominators

Subtract. Write each answer in simplest form.

1. $\frac{4}{5} - \frac{1}{5}$ — $\frac{3}{5}$
2. $\frac{7}{10} - \frac{5}{10}$ — $\frac{1}{5}$
3. $\frac{4}{7} - \frac{3}{7}$ — $\frac{1}{7}$
4. $\frac{4}{9} - \frac{1}{9}$ — $\frac{1}{3}$
5. $1\frac{5}{6} - 1\frac{2}{6}$ — $\frac{1}{2}$
6. $4\frac{7}{8} - 1\frac{4}{8}$ — $3\frac{3}{8}$
7. $2 - \frac{1}{2}$ — $1\frac{1}{2}$
8. $1 - \frac{2}{3}$ — $\frac{1}{3}$
9. $4 - \frac{1}{4}$ — $3\frac{3}{4}$

Evaluate each expression for $x = \frac{2}{8}$. Write each answer in simplest form.

10. $x + \frac{3}{8}$ — $\frac{5}{8}$
11. $x - \frac{1}{8}$ — $\frac{1}{8}$
12. $x + \frac{5}{8}$ — $\frac{7}{8}$
13. $\frac{7}{8} - x$ — $\frac{5}{8}$

Write each sum or difference in simplest form.

14. $\frac{1}{2} + \frac{1}{2}$ — 1
15. $\frac{1}{9} + \frac{3}{9}$ — $\frac{4}{9}$
16. $\frac{7}{9} - \frac{1}{9}$ — $\frac{2}{3}$
17. $\frac{2}{8} + \frac{4}{8}$ — $\frac{3}{4}$
18. $\frac{8}{11} - \frac{4}{11}$ — $\frac{4}{11}$
19. $\frac{4}{12} + \frac{4}{12}$ — $\frac{2}{3}$

20. Betty read $\frac{2}{3}$ of her book. Use pictures to model how much of her book is left for her to read, and then write your answer in simplest form.

$\frac{1}{3}$ of the book

21. Sheera ate $\frac{1}{5}$ of the pizza. Andre ate $\frac{2}{5}$ of the same pizza. Use pictures to model how much of the pizza was left over, and then write your answer in simplest form.

$\frac{2}{5}$ of the pizza

LESSON 4-8

Practice B

Adding and Subtracting with Like Denominators

Subtract. Write each answer in simplest form.

1. $1 - \frac{4}{7}$ — $\frac{3}{7}$
2. $\frac{18}{24} - \frac{10}{24}$ — $\frac{1}{3}$
3. $2\frac{2}{3} - 1\frac{1}{3}$ — $1\frac{1}{3}$
4. $8\frac{11}{13} - 5\frac{2}{13}$ — $3\frac{9}{13}$
5. $5 - 3\frac{1}{4}$ — $1\frac{3}{4}$
6. $2 - 1\frac{2}{7}$ — $\frac{5}{7}$
7. $6\frac{8}{9} - 4\frac{6}{9}$ — $2\frac{2}{9}$
8. $7\frac{4}{11} - 6\frac{3}{11}$ — $1\frac{1}{11}$
9. $10 - 5\frac{3}{5}$ — $4\frac{2}{5}$

Evaluate each expression for $x = \frac{2}{15}$. Write each answer in simplest form.

10. $x + \frac{14}{15}$ — $\frac{16}{15}$ or $1\frac{1}{15}$
11. $x - \frac{1}{15}$ — $\frac{1}{15}$
12. $\frac{13}{15} - x$ — $\frac{11}{15}$
13. $x + \frac{7}{15}$ — $\frac{3}{5}$

Write each sum or difference in simplest form.

14. $\frac{17}{21} - \frac{2}{21}$ — $\frac{5}{7}$
15. $\frac{13}{32} + \frac{9}{32}$ — $\frac{11}{16}$
16. $\frac{2}{15} + \frac{8}{15}$ — $\frac{2}{3}$
17. $27\frac{76}{100} - 14\frac{26}{100}$ — $13\frac{1}{2}$
18. $\frac{1}{15} + \frac{4}{15} + \frac{5}{15}$ — $\frac{2}{3}$
19. $\frac{9}{26} + \frac{2}{26} + \frac{5}{26}$ — $\frac{8}{13}$

20. Maria has 8 gallons of paint she wants to use in three rooms of her house. She will use $2\frac{1}{4}$ gallons in the bedroom and $1\frac{1}{4}$ gallons in the bathroom. Use pictures to model how many gallons she will have left to paint the playroom, and then write your answer in simplest form.

$4\frac{1}{2}$ gallons

LESSON 4-8

Practice C

Adding and Subtracting with Like Denominators

Subtract. Write each answer in simplest form.

1. $3 - \frac{1}{6}$ — $2\frac{5}{6}$
2. $7 - \frac{5}{7}$ — $6\frac{2}{7}$
3. $5 - \frac{3}{8}$ — $4\frac{5}{8}$
4. $4 - \frac{5}{11}$ — $3\frac{6}{11}$
5. $5\frac{8}{15} - 3\frac{3}{15}$ — $2\frac{1}{3}$
6. $8\frac{7}{8} - 4\frac{3}{8}$ — $4\frac{1}{2}$
7. $3\frac{4}{9} - 2\frac{1}{9}$ — $1\frac{1}{3}$
8. $9\frac{13}{18} - 4\frac{1}{18}$ — $5\frac{2}{3}$
9. $100 - 17\frac{9}{16}$ — $82\frac{7}{16}$

Evaluate each expression for $x = \frac{17}{36}$. Write each answer in simplest form.

10. $\frac{31}{36} - x$ — $\frac{7}{18}$
11. $\frac{1}{36} + x$ — $\frac{1}{2}$
12. $2\frac{5}{36} + x$ — $2\frac{11}{18}$
13. $5\frac{25}{36} - x$ — $5\frac{2}{9}$

Write each sum or difference in simplest form.

14. $\frac{17}{36} + \frac{10}{36}$ — $\frac{3}{4}$
15. $\frac{18}{42} + \frac{10}{42}$ — $\frac{2}{3}$
16. $\frac{88}{102} - \frac{42}{102}$ — $\frac{23}{51}$
17. $48\frac{11}{18} - 37\frac{2}{18}$ — $11\frac{1}{2}$
18. $\frac{19}{20} + \frac{16}{20} + \frac{13}{20}$ — $2\frac{2}{5}$
19. $\frac{9}{27} + \frac{3}{27} + \frac{3}{27}$ — $\frac{5}{9}$

20. Shelby practiced piano for $1\frac{1}{8}$ hours on Monday. On Tuesday she practiced for $\frac{5}{8}$ hour, and on Wednesday for $1\frac{3}{8}$ hours. Use pictures to model how many hours she should practice on Thursday if she wants to practice 5 hours this week, and then write your answer in simplest form.

    $1\frac{7}{8}$ hours

21. What are the next two terms in the following pattern: $2\frac{1}{4}$, $3\frac{1}{2}$, $4\frac{3}{4}$? Write them in simplest form.

    6 and $7\frac{1}{4}$

LESSON 4-8

Reteach

Adding and Subtracting with Like Denominators

You can use fraction strips to add like fractions.

To find $\frac{1}{8} + \frac{3}{8}$, first model the expression using $\frac{1}{8}$ fraction pieces.

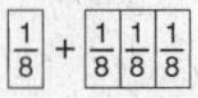

The total number of fraction pieces is 4.

So, $\frac{1}{8} + \frac{3}{8} = \frac{4}{8}$.

Then write the sum in simplest form. $\frac{4}{8} = \frac{1}{2}$.

Write each sum in simplest form.

1. $\frac{1}{4} + \frac{1}{4}$ — $\frac{1}{2}$
2. $\frac{1}{5} + \frac{3}{5}$ — $\frac{4}{5}$
3. $\frac{1}{6} + \frac{5}{6}$ — 1
4. $\frac{5}{12} + \frac{1}{12}$ — $\frac{1}{2}$

You can use fraction strips to subtract like fractions.

To find $\frac{5}{6} - \frac{1}{6}$, use fraction strips to model the first fraction.

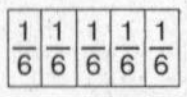

Next, take away fraction strips that represent the second fraction.

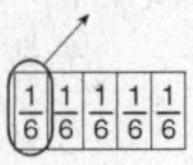

The remaining fraction strips represent the difference.

$\frac{5}{6} - \frac{1}{6} = \frac{4}{6}$

$\frac{4}{6} = \frac{2}{3}$ in simplest form.

Write each difference in simplest form.

5. $\frac{5}{8} - \frac{3}{8}$ — $\frac{1}{4}$
6. $\frac{7}{12} - \frac{5}{12}$ — $\frac{1}{6}$
7. $\frac{3}{4} - \frac{1}{4}$ — $\frac{1}{2}$
8. $\frac{9}{10} - \frac{3}{10}$ — $\frac{3}{5}$

LESSON 4-8

Challenge

Adding Equals Subtracting?

Use one expression from Box A and one from Box B to write an equation in the pans below. The two expressions must be equal to keep the scale balanced. Use each expression only once.

Possible order of students' answers are given.

A		B	
$\frac{1}{4} + \frac{2}{4}$	$\frac{1}{27} + \frac{2}{27}$	$\frac{11}{7} - \frac{6}{7}$	$\frac{11}{12} - \frac{1}{12}$
$\frac{1}{6} + \frac{4}{6}$	$\frac{2}{7} + \frac{3}{7}$	$\frac{4}{8} - \frac{1}{8}$	$1 - \frac{8}{9}$
$\frac{3}{16} + \frac{3}{16}$	$\frac{1}{5} + \frac{1}{5}$	$1 - \frac{4}{10}$	$2 - 1\frac{3}{5}$
$\frac{1}{18} + \frac{2}{18}$	$\frac{4}{15} + \frac{5}{15}$	$2 - 1\frac{1}{4}$	$\frac{5}{6} - \frac{4}{6}$

1.
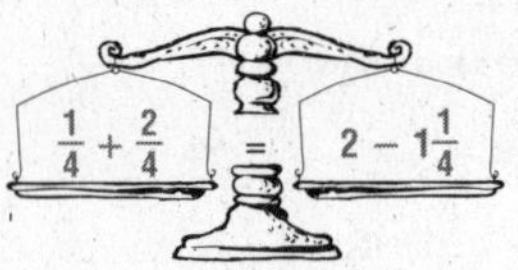

2.
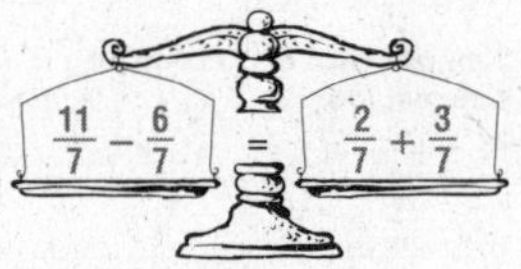

3.
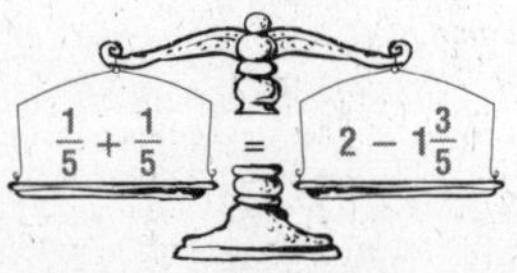

4.
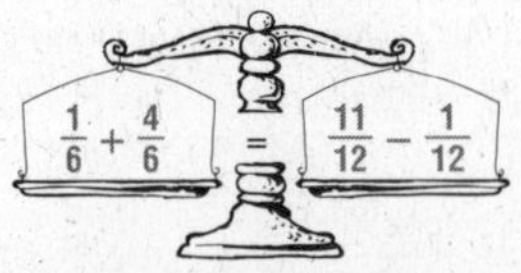

5.
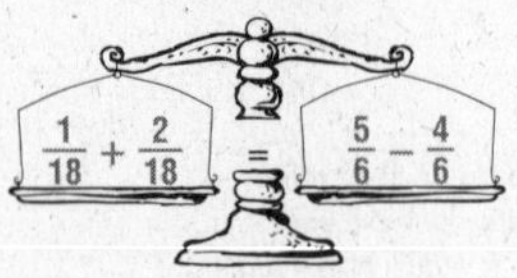

6.
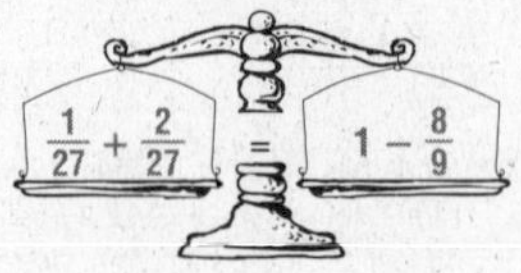

7.
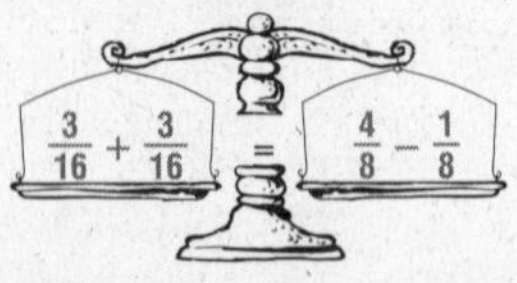

8.
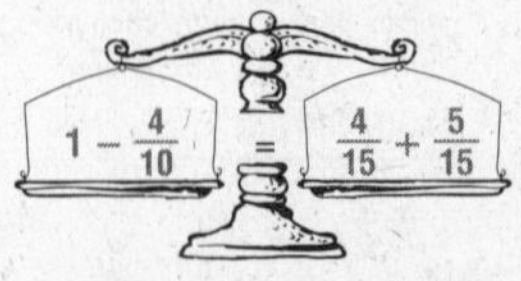

LESSON 4-8

Problem Solving

Adding and Subtracting with Like Denominators

Write the answers in simplest form.

1. About $\frac{3}{10}$ of Earth's surface is covered by land, and the rest is water. What fraction of Earth's surface is covered by water?

   $\frac{7}{10}$ of Earth's surface

2. A recipe for cookies calls for $\frac{3}{8}$ cup of chocolate chips. Tameeka wants to double the recipe. How much chocolate chips will she use?

   $\frac{3}{4}$ cup

3. In Mr. Chesterfield's science class, $\frac{2}{9}$ of the boys like his class. Three times as many girls like his science class. How many girls like Mr. Chesterfield's science class?

   $\frac{2}{3}$ of the girls

4. In the United States, $\frac{6}{50}$ of the population is left-handed men and $\frac{5}{50}$ of the population is left-handed women. What part of the population is left-handed?

   $\frac{11}{50}$

Circle the letter of the correct answer.

5. In the United States, about $\frac{1}{10}$ of the population is born with black hair, and $\frac{7}{10}$ of the population is born with brown hair. What fraction of the total population in the U.S. is born with brown or black hair?

   A $\frac{1}{10}$ C $\frac{3}{5}$
   B $\frac{1}{5}$ **(D) $\frac{4}{5}$**

6. In the United States, about $\frac{3}{20}$ of the population is born with blond hair, and $\frac{1}{20}$ of the population is born with red hair. What fraction of the total population in the U.S. is born with blond or red hair?

   F $\frac{1}{10}$ H $\frac{3}{5}$
   (G) $\frac{1}{5}$ J $\frac{4}{5}$

7. The average height for men in the United States is $5\frac{2}{3}$ feet tall. Bill is $\frac{1}{3}$ foot shorter than average. How tall is Bill?

   (A) $5\frac{1}{3}$ feet C 6 feet
   B $5\frac{3}{6}$ feet D $5\frac{1}{6}$ feet

8. The average height for women in the United States is $5\frac{1}{3}$ feet tall. Katie is $\frac{2}{3}$ foot taller than average. How tall is Katie?

   F $5\frac{1}{3}$ feet **(H) 6 feet**
   G $5\frac{3}{6}$ feet J $5\frac{1}{6}$ feet

LESSON **4-8**

Reading Strategies

Use a Visual Model

To add like fractions, add the numerators and simplify.

These rectangles show an addition problem with like fractions.

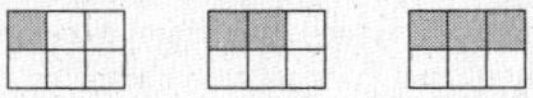

one sixth + two sixths = three sixths

$\frac{1}{6} + \frac{2}{6} = \frac{3}{6}$

$= \frac{1}{2}$

Use the model to answer the questions.

1. What is one-sixth plus one-sixth? **two-sixths**
2. What is $\frac{2}{6} + \frac{2}{6}$? $\frac{4}{6}$
3. What is three-sixths plus two-sixths? **five-sixths**

To subtract like fractions, subtract the numerators and simplify.

These rectangles show a subtraction problem with like fractions.

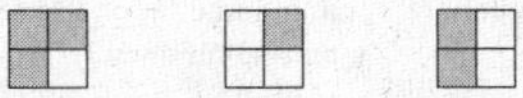

three fourths − one fourth = two fourths

$\frac{3}{4} - \frac{1}{4} = \frac{2}{4}$

$= \frac{1}{2}$

Use the rectangles to answer each question.

4. What is $\frac{2}{4} - \frac{1}{4}$? $\frac{1}{4}$
5. What is four-fourths minus one-fourth? **three-fourths**
6. What is $\frac{3}{4} - \frac{2}{4}$? $\frac{1}{4}$

LESSON **4-8**

Puzzles, Twisters & Teasers

Magic Squares

In a magic square every row, column, and diagonal has the same sum. Identify which of the squares below are magic. Then take the letters in the four corners of the squares that are magic and unscramble them to solve the riddle.

T 6	$2\frac{1}{2}$	D 5
$3\frac{1}{2}$	$4\frac{1}{2}$	$5\frac{1}{2}$
L 4	$6\frac{1}{2}$	Y 3

Sum: $13\frac{1}{2}$

Magic? **yes**

O $10\frac{1}{8}$	$8\frac{3}{8}$	A $9\frac{5}{8}$
$8\frac{7}{8}$	$9\frac{3}{8}$	$9\frac{7}{8}$
A $9\frac{1}{8}$	$10\frac{3}{8}$	L $8\frac{5}{8}$

Sum: $28\frac{1}{8}$

Magic? **yes**

B 7	$2\frac{1}{4}$	G 5
$3\frac{1}{4}$	$4\frac{1}{4}$	$5\frac{2}{4}$
N 4	$6\frac{3}{4}$	I 3

Sum: **Varies**

Magic? **no**

X 6	$5\frac{2}{3}$	S 5
$7\frac{1}{3}$	$4\frac{2}{3}$	8
K 2	$6\frac{1}{3}$	E 3

Sum: **Varies**

Magic? **no**

What did the student say when the teacher asked him if he liked studying frogs in science?

T O A D - A L L Y

LESSON **4-9**

Practice A

Estimating Fraction Sums and Differences

Round each number to 0, $\frac{1}{2}$, or 1. **Possible answers:**

1. $\frac{1}{6}$ **0**
2. $\frac{3}{7}$ $\frac{1}{2}$
3. $\frac{7}{8}$ **1**
4. $\frac{2}{5}$ $\frac{1}{2}$
5. $\frac{9}{10}$ **1**
6. $\frac{2}{15}$ **0**

Estimate each sum or difference by rounding to 0, $\frac{1}{2}$, or 1.

7. $\frac{2}{3} + \frac{3}{4}$ $1\frac{1}{2}$
8. $\frac{5}{6} - \frac{3}{5}$ $\frac{1}{2}$
9. $\frac{4}{9} + \frac{1}{8}$ $\frac{1}{2}$
10. $\frac{8}{9} - \frac{6}{7}$ **0**
11. $\frac{1}{4} + \frac{2}{3}$ **1**
12. $\frac{3}{4} - \frac{2}{3}$ $\frac{1}{2}$
13. $\frac{4}{7} + \frac{3}{5}$ **1**
14. $\frac{1}{5} + \frac{4}{9}$ $\frac{1}{2}$
15. $\frac{3}{4} - \frac{4}{7}$ $\frac{1}{2}$

Use the table for Exercises 16 and 17.

Mia's Swimming Distances

Week	Distance (mi)
1	$1\frac{1}{4}$
2	$\frac{2}{3}$
3	$1\frac{5}{6}$

16. About how far did Mia swim during week 1 and week 2 altogether? **about 2 miles**
17. About how much farther did Mia swim during week 3 than during week 1? **about 1 mile**
18. Shelley used $\frac{3}{5}$ ounce of water in her experiment. Ali used $\frac{7}{9}$ ounce of water in his experiment. Who used more water and about how much more water was used? **Ali; about $\frac{1}{2}$ ounce more**
19. Fred ran $1\frac{6}{11}$ of a mile, and then he walked $1\frac{5}{8}$ of a mile. About how many miles did Fred cover in all? **about 3 miles**

LESSON **4-9**

Practice B

Estimating Fraction Sums and Differences

Estimate each sum or difference by rounding to 0, $\frac{1}{2}$, or 1. **Possible answers:**

1. $\frac{5}{6} + \frac{3}{10}$ $1\frac{1}{2}$
2. $\frac{7}{9} - \frac{4}{5}$ **0**
3. $\frac{9}{10} - \frac{3}{7}$ $\frac{1}{2}$
4. $\frac{4}{9} + \frac{1}{4}$ **1**
5. $\frac{1}{8} + \frac{1}{6}$ **0**
6. $\frac{7}{8} - \frac{4}{5}$ **0**
7. $\frac{5}{8} + \frac{2}{7}$ **1**
8. $\frac{7}{10} + \frac{11}{12}$ $1\frac{1}{2}$
9. $\frac{8}{9} + \frac{4}{7}$ $1\frac{1}{2}$
10. $\frac{5}{11} + 1\frac{2}{7}$ **2**
11. $1\frac{6}{11} - \frac{2}{5}$ **1**
12. $4\frac{2}{7} - 1\frac{7}{9}$ $2\frac{1}{2}$

Use the table for Exercises 13–15.

Fruit Punch

Ingredient	Amount (cups)
Orange juice	$4\frac{3}{5}$
Cranberry juice	$2\frac{1}{4}$
Ginger ale	$\frac{7}{8}$

13. About how much more orange juice than ginger ale is used in the punch? **about $3\frac{1}{2}$ cups**
14. About how much juice is used in the punch? **about 7 cups**
15. About how many cups of fruit punch does this recipe make? **about 8 cups**
16. Damonte rolled the medicine ball $9\frac{3}{4}$ feet. Zachary rolled it $9\frac{7}{12}$ feet. Who rolled the medicine ball the farthest? About how much farther? **Damonte; about $\frac{1}{2}$ foot farther**
17. Sara ran $5\frac{6}{7}$ miles on Monday and $4\frac{1}{4}$ miles on Tuesday. About how many miles did she run in all during those two days? **about $10\frac{1}{2}$ miles**

LESSON 4-9
Practice C
Estimating Fraction Sums and Differences

Estimate each sum or difference by rounding to 0, $\frac{1}{2}$, or 1. Possible answers:

1. $\frac{8}{12} + \frac{5}{11}$ — 1
2. $\frac{4}{10} - \frac{2}{7}$ — 0
3. $\frac{3}{9} + \frac{3}{5}$ — 1
4. $\frac{7}{12} - \frac{8}{15}$ — 0
5. $\frac{4}{9} + \frac{7}{13}$ — 1
6. $\frac{19}{20} - \frac{7}{9}$ — 0
7. $\frac{10}{12} - \frac{5}{8} + \frac{3}{4}$ — $1\frac{1}{2}$
8. $\frac{8}{11} + \frac{9}{13} + \frac{5}{11}$ — 2
9. $\frac{6}{9} + \frac{7}{12} - \frac{2}{3}$ — $\frac{1}{2}$
10. $1\frac{2}{9} + 2\frac{3}{17}$ — 3
11. $7\frac{5}{6} - \frac{8}{9}$ — 7
12. $4\frac{8}{15} - 3\frac{2}{7}$ — $1\frac{1}{2}$

Estimate each sum or difference to compare. Write < or >.

13. $1\frac{5}{11} + 2\frac{6}{7}$ [>] 4
14. $3\frac{1}{4} - \frac{3}{5}$ [>] 2
15. $2\frac{4}{9} - 1\frac{5}{11}$ [<] 2
16. $5\frac{3}{7} + 3\frac{1}{4}$ [<] 9
17. $4\frac{6}{13} - 3\frac{2}{3}$ [<] 1
18. $3\frac{3}{10} + 2\frac{5}{14}$ [>] 4

19. Whitney's candy bag weighed $2\frac{7}{20}$ pounds. Bobby's weighed $2\frac{5}{9}$ pounds. About how much more did Bobby's candy weigh? About how much candy did they buy all together?
about the same; about 5 pounds

20. Jamie walked $3\frac{5}{7}$ miles on Monday, $4\frac{3}{11}$ miles on Tuesday, and $2\frac{5}{11}$ miles on Wednesday. About how many miles did she walk all together? Her goal was to walk 12 miles. About how much over or under her goal was she?
about $10\frac{1}{2}$ miles; about $1\frac{1}{2}$ miles under her goal

LESSON 4-9
Reteach
Estimating Fraction Sums and Differences

You can use number lines to help you estimate fraction sums and differences.

To estimate the sum of $\frac{5}{6}$ and $\frac{1}{3}$, locate each fraction on a number line. Then round each fraction to 0, $\frac{1}{2}$, or 1.

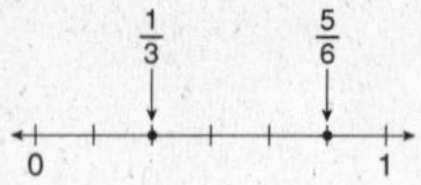

$\frac{5}{6} + \frac{1}{3} \approx$

$1 + \frac{1}{2} = 1\frac{1}{2}$

So, $\frac{5}{6} + \frac{1}{3}$ is about $1\frac{1}{2}$.

To estimate the difference between $\frac{7}{8}$ and $\frac{2}{4}$, locate each fraction on a number line. Then round each fraction to 0, $\frac{1}{2}$, or 1.

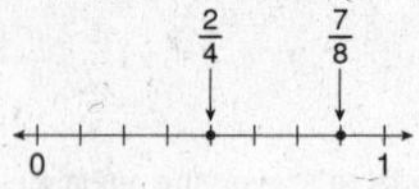

$\frac{7}{8} - \frac{2}{4} \approx$

$1 - \frac{1}{2} = \frac{1}{2}$

So, $\frac{7}{8} - \frac{2}{4}$ is about $\frac{1}{2}$.

Use the number line to round each fraction to 0, $\frac{1}{2}$, or 1 to estimate each sum or difference.

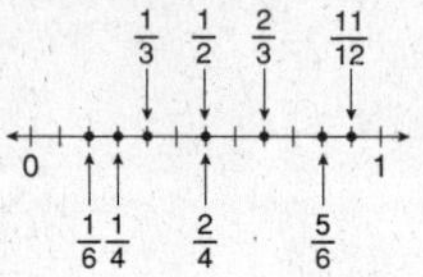

1. $\frac{5}{6} + \frac{1}{6}$ — 1
2. $\frac{11}{12} - \frac{1}{2}$ — $\frac{1}{2}$
3. $\frac{2}{3} + \frac{2}{4}$ — 1
4. $\frac{1}{4} - \frac{1}{3}$ — 0

Use the number line to round each fraction to 0, $\frac{1}{2}$, or 1 to estimate each sum or difference.

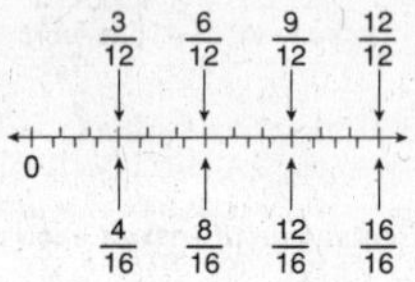

5. $\frac{7}{12} + \frac{2}{6}$ — 1
6. $\frac{5}{6} - \frac{3}{8}$ — $\frac{1}{2}$
7. $\frac{1}{4} + \frac{2}{6}$ — $\frac{1}{2}$
8. $\frac{7}{8} + \frac{14}{16}$ — 2

LESSON 4-9
Challenge
What's on the Menu?

The menu below shows the average amounts of food most Americans eat each day. Use the menu to complete the diet estimation chart that follows.

Average American Daily Menu

Dairy Products: $1\frac{2}{5}$ pounds

Processed Fruits and Vegetables: $1\frac{1}{20}$ pounds

Flour and Cereal Products: $\frac{27}{50}$ pound

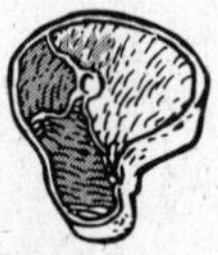
Red Meat: $\frac{9}{25}$ pound

Sweetener Products: $\frac{21}{50}$ pound

Poultry: $\frac{9}{50}$ pounds

Fresh Fruits and Vegetables: $\frac{9}{10}$ pounds

Fats and Oils: $\frac{9}{50}$ pounds

Estimated American Diet

Foods	Estimated Daily Consumption
Red meat and poultry	about $\frac{1}{2}$ pound
Fruits and vegetables	about 2 pounds
Sweetener products and fats and oils	about $\frac{1}{2}$ pound
Dairy and flour and cereal products	about 2 pounds
Total Daily Food Consumption:	about 5 pounds

LESSON 4-9
Problem Solving
Estimating Fraction Sums and Differences

Use the table to answer the questions. Possible answers:

Portland, Oregon, Average Monthly Rainfall

Month	Jan	Feb	Mar	Apr	May	Jun	Jul	Aug	Sep	Oct	Nov	Dec
Rain (in.)	$5\frac{2}{5}$	$3\frac{9}{10}$	$3\frac{3}{5}$	$2\frac{2}{5}$	$2\frac{1}{10}$	$1\frac{1}{2}$	$\frac{3}{5}$	$1\frac{1}{10}$	$1\frac{4}{5}$	$2\frac{7}{10}$	$5\frac{3}{10}$	$6\frac{1}{10}$

1. About how much does it rain in Portland in January and February combined?
about $9\frac{1}{2}$ inches

2. About how much more does it rain in Portland in October than in September?
about 1 inch more

3. In most years, about how much rain does Portland receive from May through July?
about 4 inches

4. What is the difference between Portland's average rainfall in March and May?
about $1\frac{1}{2}$ inches

Circle the letter of the correct answer.

5. What is the difference in rainfall between Portland's rainiest and driest months?
   A about $2\frac{1}{2}$ inches
   (B) about 5 inches
   C about $6\frac{1}{2}$ inches
   D about $7\frac{1}{2}$ inches

6. About how much rain does Portland receive in most years all together?
   F about $25\frac{1}{2}$ inches
   G about $30\frac{1}{2}$ inches
   H about $32\frac{1}{2}$ inches
   (J) about $36\frac{1}{2}$ inches

7. About how much rain does Portland receive during its three rainiest months all together?
   (A) about 17 inches
   B about 16 inches
   C about 18 inches
   D about 15 inches

8. In which month in Portland can you expect about $\frac{1}{2}$ inch less rainfall than in June?
   F May
   G July
   H September
   (J) August

LESSON 4-9 Reading Strategies
Use a Graphic Aid

When you don't need exact values of fractions, you can use a number line to help you estimate the values by rounding.

Fractions Close to 1
The number line shows that the fractions $\frac{5}{6}$ and $\frac{7}{8}$ are both close to 1. When the numerator and denominator of a fraction are close to the same value, round to 1.

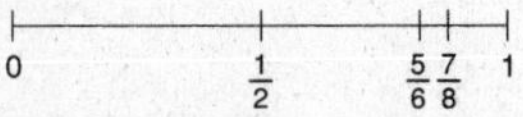

Fractions Close to $\frac{1}{2}$
The number line shows fractions such as $\frac{6}{13}$, $\frac{9}{20}$, and $\frac{12}{23}$ that are close to $\frac{1}{2}$. When the numerator is about half the value of the denominator, round to $\frac{1}{2}$.

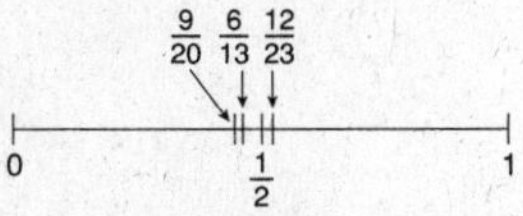

Fractions Close to 0
The number line shows fractions such as $\frac{1}{15}$, $\frac{6}{50}$, and $\frac{2}{25}$ that are close to 0. When the numerator is much less than the denominator, round to 0.

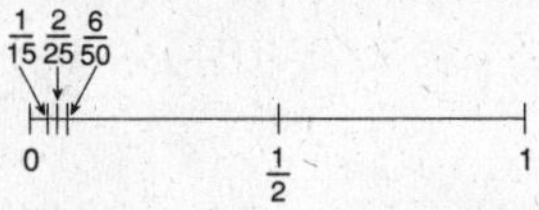

Estimate the value of each fraction. Write *close to 0, close to $\frac{1}{2}$,* or *close to 1.*

1. $\frac{12}{25}$ close to $\frac{1}{2}$
2. $\frac{1}{17}$ close to 0
3. $\frac{8}{9}$ close to 1
4. $\frac{9}{10}$ close to 1

LESSON 4-9 Puzzles, Twisters & Teasers
Call to Arms

Estimate the sum or difference by rounding each fraction or mixed number. Circle your estimated answers.

To solve the riddle, fill in the blanks with the matching letters.

1. About how much is $\frac{7}{8} + \frac{5}{6}$?

   (R) 2 U 3 W 4

2. About how much is $6\frac{2}{3} - 2\frac{1}{4}$?

   K 3 (D) 4 S 5

3. About how much is $2\frac{5}{7} + 3\frac{1}{4}$?

   H 5 (E) 6 C 7

4. About how much is $8\frac{1}{4} - 7\frac{5}{6}$?

   B 2 T 1 (A) 0

5. About how much is $4\frac{3}{8} + \frac{4}{7}$?

   (M) 5 F 6 P 4

How does an octopus go to war? A R M E D
(4 1 5 3 2)